SHARP is a knife-skills class in book form and an introduction to the best knives you can buy from all over the world. From premier knife purveyor and go-to knives and sharpening expert Josh Donald of Bernal Cutlery this is the book that teaches

- **what knives to buy**
- **how to care for your knives**
- **how to sharpen your knives**
- **dozens of precise cuts, from simple dicing to fancier oblique shapes**

All of the cuts are practiced through recipes from leading chefs, including Stuart Brioza of State Bird Provisions, Melissa Perello of Frances, and Nick Balla of Duna. From a simple salsa verde to an exquisite noodle soup, here are dozens of delicious ways to perfect your skills.

Josh travels across the globe to meet the knife makers he features in his shop, to learn from the source the best ways to handle and sharpen knives. In *Sharp*, he passes this knowledge to you. Join him on a visual journey through hundreds of photographs of people, forges, metal shops, tools, and knives in Japan, France, Germany, and the United States. Explore a range of vintage knife types, as well as the history of knife making, from earliest forging and metallurgy to the limited edition, much-coveted knives of today.

All of the techniques in this book are clearly illustrated with step-by-step photography, from making cuts to sharpening methods. This book is packed with beautiful, helpful information to hone your knife skills and transform you into a knife expert.

SHARP

SHARP

The
Definitive
Introduction
to Knives,
Sharpening,
and Cutting
Techniques,
with Recipes
from Great
Chefs

Josh Donald with **Molly Gore**

Photographs by Molly DeCoudreaux

CHRONICLE BOOKS
SAN FRANCISCO

Library of Congress Cataloging-in-Publication Data

Names: Donald, Josh, author.
Title: Sharp / by Josh Donald.
Description: San Francisco : Chronicle Books, 2018.
Identifiers: LCCN 2017022740 | ISBN 9781452163062
   (hc : alk. paper)
Subjects: LCSH: Knives. | Cutting.
Classification: LCC TS380 .D675 2018 | DDC 621.9/32—dc23
   LC record available at https://lccn.loc.gov/2017022740

Manufactured in China.

MIX
Paper from
responsible sources
FSC™ C008047
www.fsc.org

Designed by Toni Tajima and Vanessa Dina
Typesetting by Frank Brayton

10 9 8 7 6 5 4 3

Chronicle books and gifts are available at special quantity discounts to corporations, professional associations, literacy programs, and other organizations. For details and discount information, please contact our premiums department at corporatesales@chroniclebooks.com or at 1-800-759-0190.

Chronicle Books LLC
680 Second Street
San Francisco, California 94107
www.chroniclebooks.com

There are so many people who have helped to create this book, who've supported our mission at Bernal Cutlery and who make me love the business I'm in. They include:

Kelly Kozak, who refuses to let me be a pessimist and has my back always. The staff at Bernal Cutlery, who not only contributed to this book but kept the gears turning and without whom I would have been checking in orders and sharpening knives instead of writing this. Molly DeCoudreaux, who has always given a fresh perspective on what we do through her steely eye and steady lens. Consigliere Liam Passmore, who introduced me to Danielle Svetcov, who has been a fountain of sage advice and advocacy. Molly Gore, whose literary skills and hard work shaped and honed my rough-forged run-ons into a fluid, readable form. Chronicle Books, for giving us a chance and creative license. All the cooks in SF who contributed their knife skills and recipes; we wish we could've fit everyone in!

All those in Japan who were so generous with their time: Ashi-san at Ashi Hamono, who helped us take a deeper look into the history of the gyuto and the roots of modern Japanese knife making. Kawamura-san, Yoshida-san, Koda-san, and the good people at Sakai Kikumori Kawamura Hamono who introduced us to their network of craftspeople, for a deeper look into traditional knife making in Sakai. In Kyoto, Dobashi-san of Maruoyama mine, who welcomed us into his shop Totoriya and climbed the steep hills of Kameoka with us for a tour. Manfred and Sachiko Ikeda, who introduced us to the amazing artisans of Tsubame-Sanjo and helped organize the Niigata end of our travels. Iwasaki-san and Mizuochi-san, gracious hosts who shared a deeper context and understanding of traditional smithing and sharpening. Tsukasa Hinoura-san, who took time out of his busy schedule to share his knowledge and perspective on knife making and the inner workings of traditional Japanese blacksmithing. Mutsumi Hinoura-san, who also generously shared his experience. Yamamoto-san of Yoshikane, who brought us through their unique production process and gave us a new look at knife making. Wakui-san, who was also very generous with his time and techniques. In Europe, a big thank you to Giselheid Herder-Scholz, who was invaluable in making sure our Solingen research was thorough and blew my mind with her family collection of Windmuhlenmesser knives. In France, an enormous thank you to the Bournilhas Family: Alain, Philippe, and Christine, amazing hosts who shared the hidden world of Thiers, the rich history of K Sabatier, including massive ledger books, and their collection of knives.

Finally Shihan Prull, who generously took me through all the mechanics of forging and heat treating and enriched my understanding of knife making.

Contents

Introduction

In this book, you'll learn about knives from the West and Japan and from the moments in time when those worlds crashed and then melted into each other. These are the knives you'll find in most professional kitchens and, chances are, in your own kitchen, too. You'll learn to look at a knife and see right to its material roots— from the ore to the furnace to the polishing wheel—and beyond them to the story behind its footprint and the way it feels when you cut with it.

This book holds what I now know after ten years of jumping down rabbit holes and tracing everything we sell at Bernal Cutlery to its origin. Along the way, I learned how to work a Japanese whetstone (first badly, then well) and how a sharp edge and cutting angle affect the flavor of food. I found out how Commodore Perry's gunboats opened the door for the arrival of beef to Japan, and how single-celled organisms that fell to the seafloor of Pangaea hundreds of millions of years ago can turn an onion to butter under the edge of a Japanese steel knife. In France, Germany, and Japan, in the forges and small workshops where our favorite knives are made, I saw how finishing a blade by hand against a grinding wheel really makes it sing on the cutting board.

You'll find all of that in these pages, as well as a practical guide to putting knives to use with techniques and recipes. In chapter 1, I begin with the raw materials because that's where all knives were born, and then move through the history of smelting in Japan and the West. In chapter 2, I go deep into the history of knives in Europe, from the first restaurant in the mid-eighteenth century through the rise and fall of Europe's preeminence in knife making. In chapter 3, the focus turns to Japan, to the ancient furnace, black iron sands, and prehistoric mud that drove Japanese knife making for centuries, and to the craftspeople who keep that legacy intact today.

The second half of the book is devoted to content that helps you use your knives to their best advantage. I begin with a sharpening lesson, exactly the way I've learned to teach at Bernal Cutlery after ten years of ironing out the kinks. You'll find a practical guide to knife skills and recipes from chefs who frequent the cutlery, including Traci Des Jardins of Jardinière, Stuart Brioza of State Bird Provisions, and Loretta Keller, formerly of Coco500, all converts to my favorite knives who love them as much as I do.

All of this has been bundled into a single book because the way a knife performs has everything to do with how it came to be. The story of knives on this planet is a story of human beings and of movement, conquests, and trade. When you take a sharp, hand-ground *gyuto* to a carrot, it's like taking a hot knife to cold butter, and that has everything to do with the development of the hard, fine steel along its edge that evolved to suit the fine-grained sharpening stones that have been mined from a band of rock outside of Kyoto for centuries. And the design of the *gyuto*, the downward sloping spine and wide footprint, came about to suit the food that arrived in Japanese kitchens when the United States forced that country's borders open.

I find it difficult to talk about a knife without talking about the world around it. Following the steel brings you to Japan, but chasing the footprint that inspired it lands you right in the middle of eighteenth-century Paris, when the restaurant was born. Knives like the *gyuto* or the small petty knife, which take their form from the West but their feel from Japan, inevitably lead us to those crashing and melting moments in history that brought Japanese and Western culture together outside and inside the kitchen.

There are other ways to trace a knife. I've found that following the cutting feel of a Western or Japanese knife will take you back to the person who made it and the traditions that maker has kept. Today's Japanese knifesmiths inherited a thousand-year-old blacksmithing tradition, and they bring a millennium of practice manipulating steel into the knives I buy from Sanjō and Sakai. The broad majority of the knives we sell are made using skilled handwork—and that's what makes the knives sing: shaped to just the right convexity against the walrus skin on a polishing wheel, or forge welded with just the right balance of hard and soft steels. There's no substitute for skilled hands, and these knives are in our shop because Sakai, Sanjō, Solingen (German), and Thiers (French) have protected their artisans and heritage so well. These towns are not the only places where old-style, preindustrial smithing techniques survive, but through them we can tell the story of what makes the heart of Bernal Cutlery beat: artistry and traditional techniques. This knowledge survives because certain people are working to keep it alive, and Bernal Cutlery is lucky to have enduring relationships with them.

After many years of selling these knives, I found myself in Japan on a pilgrimage to those workshops, to meet the makers in person so that I might better understand the forging, grinding, and sharpening techniques that make a knife sing. I returned two years later to do some research for this book, and on a cold April morning, I met Kazuomi Yamamoto in a little meeting room above the factory floor at the Yoshikane forge. We drank green tea, and Yamamoto-san was serious and direct, if amused, with me when I dogged him (politely) for his secrets and for answers about his family's style of knife making. After nearly a century in the business, he left me with only two gems: *sessa takuma* and *shoshin wasuru bekarazu*.

The Japanese characters that make up *sessa takuma* translate roughly to "cut, shine, rock, and grind," which suggests a friendly competition in which we study others' skills to improve our own and to pass that knowledge on. The characters themselves look a little like rocks tumbling together to smooth their edges, refining one another in order to refine themselves. Without their fellow tumblers, none of them would roll smoothly. The idea is that knife makers ought to study each others' skills to improve their own and share what they know. *Shoshin wasuru bekarazu* tells us to remember the spirit and youth of the beginner's mind and that learning never ends. By embodying both philosophies, a craftsperson opens space for skills to develop and egos to soften. And that's the space where exceptional knives are made.

My personal growth as a sharpener has been driven by working with cooks and seeing knives come back into the shop in various conditions for maintenance. As a sharpener and teacher, I enjoy nothing more than watching someone create his or her first good edge. Those initial edges bring me back to my own beginning, to my yearning to learn, to my *shoshin wasaru bekarazu*. There is still so much to know if I remember to keep experimenting and asking questions.

I offer this book in that spirit. The craftspeople you will meet in these pages have mastered skills that will take me a lifetime (and then some) to understand fully. At no point should you get the impression that I have anything more than a rudimentary comprehension of what they know. I don't speak Japanese, I am not a metallurgist, I don't have a time machine, and I am a middling cook. I am always most inspired by the variety of people and the new skills that flow through the knife shop on a good day, reminding me of how many things I want to learn. I hope you will find in these pages a few techniques to add to your wheelhouse and that you get half as much out of reading this book as I have gotten out of working on it.

HOW I CAME TO KNOW AND LOVE KNIVES

Knives have been part of my life since the beginning—or at least since one Sunday morning in 1978. I was five years old and woke up early. I remember thinking that the house and all of its contraband were mine until my mother woke up. I could plunder the kitchen drawers stuffed with arcane and sinister implements, gobble forbidden chocolate chips, and prowl through the contents of any shelf within reach.

It was in the stale wooden confines of an almost-empty dresser that I found it: a small, gold-hued penknife. It winked at me. Bonanza! My hand closed around it, and just like that, I had my very first knife. It was little, yeah, but it gave me the courage to step up to the bigger, equally neglected blades that populated our house, and eventually it fell out of favor when I got my hands on those. Soon after, I liberated the giant carbon-steel chef knife collecting sticky dust in the back of a kitchen drawer, right behind the orange-plastic Parmesan cheese grater.

I took it and stepped out of the house and into the dusty, dry grass of our overgrown yard, the sound of freight-train brakes echoing from the tracks near the Los Angeles River. As I carved tunnels in the dense brush, cleaving branches and vines with the gray carbon-steel blade, I knew only two things: One, I could get in a lot of trouble. Two, I was having the time of my life.

The moment I heard my mother moving around in the house, I stashed the knife in a tunnel I'd carved into the center of a massive honeysuckle bush. Over the next several weekends, my backyard visits became a ritual, until one morning when I unexpectedly heard the sound of my mom's voice behind me in the yard. In a panic, I threw the chef knife into a shrub. Despite my best efforts, I never found it again.

Over the next couple of years, new knives came into my life. When I was six, my dad let me cut vegetables for dinner at his new apartment in West Hollywood: I remember standing at the cutting board with one of his knives—it was either a Chicago cutlery boning knife or the mighty Ginsu—in my hand. I can't remember what I cut or, for that matter, cutting myself. But the heady feeling of being entrusted with a knife? I remember that like it happened yesterday.

At the end of my first-grade year, Dad gave me an Imperial Barlow pocket-knife with a brown-plastic shell handle, stamped steel liners, and carbon-steel blades. Millions of Barlows were made by Imperial—you'll always find them in a box at a flea market or junk shop—and they all seem to get a chip on the butt end of the handle like mine did. I loved that knife and saw it as a consolation prize for what had been a rotten school year under the tutelage of one Mrs. Trope, a woman who looked like Hitler in a beehive.

But the Barlow paled in comparison to the Buck knife I got for my eighth birthday. The Buck was much bigger than the Barlow and had a blocky handle with two brass bolsters and rosewood scales. I took it everywhere: it was such a constant presence that, during the summer, it wore a hole in the pocket of my blue corduroy OP shorts. Eventually the blade became so dull that its edge grew round and shiny, and its modest 2-in/5-cm sharpening stone, which was fragrant with sharp-smelling oil, was effective only for cutting my fingers and for making a mess on the kitchen table. I'm not sure when I eventually lost that Buck knife, but it was probably once I was a teenager, long after its potency had been eclipsed.

I met my first Japanese knife and sharpening stone when I was nineteen. The year before, I'd moved to San Francisco, where I was the very portrait of the artist as a young man. Back then, you could afford to do that in San Francisco. After settling into my new apartment in the Mission District (which, as fate would have it, was literally around the corner from where I would open my second Bernal Cutlery storefront two decades later), I went to a large hardware store in Japantown to buy some supplies. I wasn't planning to buy much, as money was tight, even though my monthly rent was only $240. But after finding a small stash of rustic hand-forged Japanese knives, I decided to skip buying weed in Dolores Park that week and walked out of the store with an eighteen-dollar carbon-steel *santoku* knife.

I loved that knife, but after several months of use, it was no longer the little badass that sang through vegetables on their way to becoming bad nineties vegetarian stir-fries. It sat untouched, one side rusted over, in a kitchen drawer. So I went back to the Japantown hardware store and sought guidance and a sharpening stone. The former was in short supply: the man stationed behind the handmade chisels simply pointed in the direction of a display case and said, "Stones are over there." He then walked away, leaving me to face rows of boxes stamped with kanji, grit numbers, and price tags. I didn't know much about Japanese stones, but I figured the middle numbers would be the best choice—neither too coarse nor too fine. An hour or so later, I was back in the Mission with my new red King medium-grit stone, ready to experiment.

In 1994, I was going it alone with my medium-grit stone, making a red, muddy mess next to the sink in my kitchen, not having a grasp of the basic techniques or mechanics of sharpening. But there was something so satisfying about finally getting a glimpse of the edge I wanted that it made me want to learn how to do it right.

In 2005, I was sitting at the kitchen table in the Bernal Heights apartment I shared with my wife, Kelly, and our nine-month-old son, Charlie. I was freshly unemployed, having been laid off from a decorative-bronze hardware company. It had been part of a short-lived career switch and proved to be a nightmare job that, after a scant four months, ended with tears of joy. At the time, Kelly's work as a freelance photography assistant was taking off, but the length of her jobs was unpredictable, and we still needed to find a way to buy groceries reliably. As we talked about what I could do to bring in some cash, I sharpened our kitchen knives with my now well-worn King sharpening stone.

As I sat there sharpening our knives, I realized that even though I had never mastered it, sharpening knives was something I enjoyed. And I knew that what I lacked in skill, I made up for in determination. And so, with the forty-dollar purchase of a second Japanese stone and a handful of flyers printed on our home computer, Bernal Cutlery was born.

I set up my workbench, an old school desk, on the sloping floor of our apartment's utility room. My first orders were a mixed bag: not only had some of the knives never been sharpened, but they'd been bought as part of one of those fifty-dollar sets of a dozen, and like a lot of mass-produced knives, they were resistant to sharpening. I had to work like hell to get an edge on them, which meant my wrists ached and, given the original condition of the knives, I was making about $5.25 an hour. I'd get an order started while Charlie napped, and then later, we'd go out together to pick up more, the bundles of knives weighing down the webbing on the underside of Charlie's stroller.

It was around this time that I met Frank at the flea market near my house. An older gentleman, he looked a little like Charles Bukowski, used phrases like "you're still shitting yellow," and had an extensive knowledge of Persian rugs and of American arts and crafts furniture. He also had a box full of used kitchen knives priced mostly at five or ten dollars. To a casual observer, it looked like a box full of junk, and at first, I had no idea what I was looking at. But I soon discovered that nestled under the unloved (and unwashed) dollar-store knives were elegant, ebony-handled antique French Sabatiers, robust but graceful German blades, and knives made by hand along the leafy rivers of Massachusetts. I started buying those knives and then selling them at restaurants and on eBay, in the process learning which ones would fetch ten dollars and which ones would bring me twenty times that amount. I realized I was onto something great when an older French cook lost his poker face while appraising a hand-forged circa-1900 French fillet knife stamped with a trademark of a fireman running with a ladder. I was selling it for somewhere between fifty and one hundred dollars too little.

Still, I didn't know quite how good I had it until a few years later when Frank passed away and the boxes of knives disappeared. I continued to find my fair share of antique knives by showing up at the flea market every Sunday at 5:30 a.m. By this point, I was experienced enough to spot the good stuff in faint light without cutting my fingers too often. With Charlie now in preschool, I had a lot more time for orders, and my sharpening had improved greatly, thanks in part to a small whetstone wheel grinder that took much of the manual labor out of the task.

Back then, I was going around offering free sharpening to restaurants to introduce them to my service. I sculpted a lot of cheap, banana-shaped blades attached to dirty white-plastic handles back into the shape of a chef knife, but most places never contacted me again or balked at the five dollars I was asking for knives with blades over 6 in/15 cm long.

I can't exactly remember which professional kitchen gave me my first order. It was either Blue Plate, a then relatively new restaurant on Mission Street, or Moki's, a Japanese place in Bernal Heights. But I do have a clear memory of meeting Cory Obenour, Blue Plate's chef and co-owner, when he pulled up in his car on Cortland Avenue in front of my apartment and unloaded bundles of knives from the trunk. It was an exchange that bore a passing resemblance to the gun and drug deals I'd occasionally see from my apartment window before realtors started calling Bernal Heights "Bernal Village." Cory, whom I found instantly likable, looked like my junior high school vice principal, but instead of collecting golf umbrellas and prim Italian suits, he looked like he rode a skateboard in empty swimming pools and worked the grill station with equal panache.

Cory had a Swedish carbon-steel Misono knife with a deep gray patina and a dragon etched on the side. I stayed up late finishing his order, and I still remember how proud I was of the way the bright polished bevels of the Misono contrasted against its dark patina. The next morning, I arrived

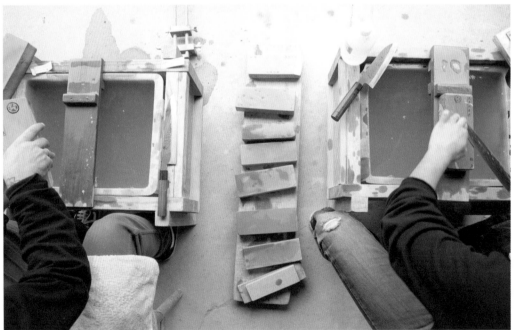

at Blue Plate with a bouquet of knives wrapped in several newspapers held fast with masking tape. I worried that I'd get a phone call saying the work sucked, but Cory was happy—stoked, even—and instead of an angry phone call, I got a nice check.

The knife I sharpened from Moki's was as life altering for me as the order from Blue Plate had been: it was a traditional Japanese sashimi knife called a *yanagi* (also known as a *yanagiba*). The *yanagi* is sharpened on one side, more like a chisel than a classic Western knife. And whereas you might remove 2 to 3 mm/$\frac{1}{16}$ in of metal along the bevel of a Western knife, it might be 20 to 30 mm/$\frac{3}{4}$ to $1\frac{3}{16}$ in or more on a *yanagi*. The one I brought home from Moki's was my first, and I wasn't charging for it. It was also significantly misshapen. I started to sharpen it with a coarse diamond stone, and several hours later had taken off only a quarter of the metal I needed to remove to reshape the edge, much less polish it. As I worked through the day into the evening, the ghosts of the ten thousand fish that it had rendered into sashimi awoke and released their fragrance onto my hands, which smelled for the next day or two. But I finally got the *yanagi* sharpened, and from then on I was entrusted with Moki's good knives.

Up to this point, I had been largely ignorant of many of the nuances of Japanese knives and steels. It was when I first used a secondhand factory-made 240-mm/$9\frac{1}{2}$-in *gyuto* to slice through a cabbage that a lightbulb switched on in my head. It was effortless and completely different from what I had experienced with the thicker Western blades I had been using. The only knives that came close were some of the thin-bladed, finely forged old Sabatiers that I'd found in nearly new condition. (For the purposes of this book, I'll refer to the size of Japanese knives in millimeters and western knives in inches, because this is how they're known among the cooks who use them.)

Not long after my lightbulb moment with the *gyuto*, Kelly, Charlie, and I packed up and moved around the corner to a new apartment on the ground floor of a building that had once housed a nineteenth-century grocery store. It was an eventful time. Right after we moved, our second son, Henry, was born, and not long after that, so was the third phase of my business: I set up a workshop in our front room, outfitted it with a nineteenth-century oak display case that Kelly found for two hundred dollars on Craigslist (it had a matching sister case, but we didn't have the matching two hundred dollars), and opened my unofficial, invitation-only shop. I held monthly sharpening classes on a folding table and bought a dozen or so relatively inexpensive Western-style carbon-steel Japanese chef knives from a source I'd found in the United States to stock the display case and sell. I continued sharpening knives, but now I strapped Henry to my back and used his stroller to cart the knives to customers and boxes of eBay sales to the post office. With the stroller loaded up and Henry on my back, we looked like some latter-day urban version of the Joad family.

As word got around among both chefs and home cooks, my business grew, and in 2010, Bernal Cutlery entered its fourth incarnation when a local landlord asked if I'd like to rent a stall in a small culinary marketplace on Cortland Avenue, in the heart of Bernal Heights. At 90 square feet/8.5 sq m, the stall was basically the size of a dashboard, but it allowed me to finally take Bernal Cutlery out of our family apartment and give it its first real storefront. That stall was also where, after about a year in business, I began directly importing Japanese knives.

In the course of my research on knife makers in Sakai City, the center of traditional knife manufacturing in Japan, I came across the website for Ashi Hamono, a knife company owned by two brothers. I liked that one of them was a sculptor, so I contacted the brothers, and they agreed to sell to me. They, in turn, liked that I did Japanese-style sharpening and felt that I'd be able to give the knives proper care after they were purchased. I didn't know it then, but I'd just scored big: because of its small size, Ashi Hamono stopped taking new wholesale customers after me. I had gotten in just in time.

Ashi Hamono makes light, thin, single-steel (generally not forged) Swedish stainless-steel and Japanese carbon-steel knives with Western or Japanese handles in several dozen styles and sizes. Deciding which of the company's knives to order was agonizing, particularly given my shopping budget of less than $2,000. I remember the day the knives arrived, their boxes sealed with duct tape the color of a Band-Aid. That tape was my first indication that these knives were different from the ones I'd sourced domestically: their boxes were light and clean, and their blades were less than 2 mm/$\frac{1}{16}$ in thick. I was besotted. Fortunately, so were my customers. Those knives cut like no other knife I had known and were as much of a revelation as the first *gyuto* I had used. And compared to the mind-blowing ease of the Ashi, that *gyuto* now felt like a sharpened screwdriver. It didn't take long for

the Ashi knives to gather a Bay Area following: all I needed to do to win converts was to set a carrot next to a knife.

My store got busy and began to outgrow its marketplace stall. But in 2013, fate once again intervened in the form of an empty—and surprisingly affordable—storefront in the Mission District. It wasn't big (less than 600 square feet/58 sq m) but more space meant room for more sharpening and more inventory, and four years later, Bernal Cutlery was once again running out of space. We took over a small place around the corner so I could teach sharpening classes to more than two people at a time, fill orders, and upgrade our inventory space from our hobbit-size office, which we climb a ladder to get to.

Scores of professional chefs come to Bernal Cutlery to buy new knives and have old ones sharpened, but so do home cooks. It is for the latter in particular that the shop hosts an ongoing series of knife-skills classes that teach everything from how to dice an onion to how to fillet a fish. That Mission storefront is my little corner of paradise: it's where I get to practice a craft and share my excitement with the rest of the world.

Chapter 1

A History of Edged Tools in the West and East

For our purposes, this story starts a few million years ago, when all we had were rocks. Knives ride the coattails (or bearskins, loincloths, or whatever) of human history, evolving as people move, meet, and discover new materials.

STONE

You might be wondering why, in this book that promised to teach you about kitchen knives, we're starting in the Stone Age? All I can say is that the oldest cutting tools chipped from rocks were made a dumbfounding 3.3 million years ago, hundreds of thousands of years before the earliest humans had evolved. Those first tools—jagged shards of stone discovered near Lake Turkana in Kenya—were unsophisticated, but they did the job. And most important, along with fire, they drove our ability to cook, grow bigger brains, and self-domesticate. The capacity to cut things and to cook changed not only how we ate but also our physiology.

Crude as they were, early stone tools could take an amazingly sharp edge. Any stone with a sufficiently fine crystal structure can fracture along a razor-sharp edge, even sharper than steel. But stone is brittle, and rocks are only so big (before they become too heavy to lift), so the size of tools back then was limited. It seems—judging by the evidence that archaeologists have chiseled from sites like Lake Turkana—that these tools evolved slowly, in fits and starts, until the *Homo sapiens* showed up two hundred thousand years ago and sparked an explosive change in style and functionality. As time went on, hunting-and-gathering societies evolved into an agricultural people, and the shape and functionality of their tools continued to change. Then, after a couple of million years, the materials changed, too.

BRONZE

As smelting technology evolved, the Middle East and then the West adopted copper as tool material. Fine deposits of copper are spread throughout the earth, but it's rare to find pure chunks. Instead, like most metals, copper is usually mixed with other elements embedded in some kind of host rock that we call ore, and to use it, we smelt the copper out. Copper relics unearthed in Israel's Timna Valley date back nine thousand years, which tells us that humans have been smelting for a long time.

To separate copper from its ore, the ore is heated to a high temperature to induce a chemical reaction that causes the pure copper to bind together and melt from the rock. Luckily for our ancestors, copper could be smelted at a lower temperature than other metals (around 1,984°F/1,084°C)—easy enough to reach with a pottery kiln. Kilns were widespread nine thousand years ago, and anyone who used them would have had access to copper, as long as they had the ore.

Even in the pre-Columbian Americas, where metalworking never truly eclipsed the use of stone for edged tools, evidence of copper smelting exists. The Andes and a swath of Wisconsin were, and still are, host to large deposits of native copper veins and nuggets. Copper has always been easy to find, but somewhat soft and impractical. It could be hammered into shape to create artworks and tools that didn't require a high degree of hardness, but it can't be cast in its pure form, which limits the number of shapes it can take. And because it could not take a very keen edge, it never fully replaced stone for blades.

Bronze, on the other hand, could. Eventually it was discovered that other metals could be added to copper to make casting easier, and that adding arsenic or, later, tin, yielded bronze: harder than copper, far tougher than stone, and easily cast into a wide variety of shapes that made new tools, notably edged ones, possible.

IRON

After a couple of thousand years, the Bronze Age faded into the Iron Age, which most scholars agree had reached the majority of the world by 1200 B.C.E. Iron is the fourth most common element on Earth, but due to its love of oxygen it is never found in its native, solid, unoxidized form outside of meteoric iron, which is an iron-and-nickel alloy that has been mined for thousands of years from the rare meteors on the earth's surface. Iron might be the most widespread, but it melts at 2,800°F/1,538°C, roughly 900°F/500°C hotter than the melting point of copper and more than six times hotter than that of tin. A pottery kiln is too weak to reach those temperatures, so smelting iron and making use of it as tool material didn't happen until stronger furnaces evolved. This happened at different times across the world, and in some regions, bronze smelting was skipped entirely and people went straight from chipping stone to smelting iron.

But making use of iron requires more than just smelting it. Impurities must be driven out to make it consistent, and the carbon levels refined to control hardness. The impure low-carbon iron produced by the first furnaces was no harder than bronze, but the ore was much easier to find than copper and tin, making iron much more popular as tool material. Making practical tools was a matter of manipulating the carbon levels, and as the technology to do that became more precise and sophisticated, a rudimentary precursor to steel evolved. But before that, there was only smelting and hammering.

Iron-smelting furnaces are not like kilns. The earliest designs involved a kind of cylindrical earthen enclosure with a layer of charcoal at the bottom and a tuyere, or air pipe, attached to a set of bellows. The bellows pumped air through the tuyere, feeding the smoldering coals and heating the furnace. As the iron ore heated up, it released oxygen and melted into a spongy mass mixed with slag, the glassy, nonmetallic remains of the ore.

At this stage, a blacksmith would pull the glowing spongy mass from the bottom of the furnace and hammer it to drive out the ash and slag until all that was left was somewhat pure wrought iron, which was tough, ductile, and low in carbon (between 0.02 and 0.08 percent). As smelting technology evolved, iron replaced bronze as the material of choice for edged tools, and the world changed. Economies that relied on bronze currency crashed along with the empires that depended on them, introducing a period of massive restructuring. Because iron was more widespread than copper and tin, the availability of metal tools increased dramatically. Before iron, access to tools depended on infrastructure, on trade relationships with ore-rich regions, and on the knowledge and labor to mine and smelt local ores. But now, almost everyone had access to tools.

Furnaces continued to evolve, and so did the iron they made. In the West during the Middle Ages, the hand-pumped bellows and primitive furnaces were replaced by tall, narrow blast furnaces powered by waterwheels that channeled larger volumes of air through the charcoal and ore. In a blast furnace, the iron ore reached 1,300°F/704°C, which was so hot it began to absorb carbon that dropped the melting point. The molten iron then flowed into troughs that branched into a pattern of ingots, which, in the imagination of some medieval smelter worker, must have looked like piglets nursing from their mother. It is still called "pig iron."

Pig iron can be cast, just like bronze, and when it is mixed with alloys like silicon and purified to some degree, it hardens into cast iron, which is a popular material for pans, thanks to its excellent heat retention (but has a tendency to break, which you'll know if you've ever dropped a cast-iron pan). Cast iron has at least 2 percent carbon, less than pig iron (which is 4 percent or more), but it's still too brittle to make a useful tool.

Up to this point in our story, we've met two types of iron—wrought and pig—and each has its flaws. Wrought iron is tough but malleable, low in carbon, and good for shock-bearing tools like horseshoes. Pig iron is brittle, moldable, high in carbon, and easy to crack. To make something with more utility, that's hard but durable, we need an iron with a carbon content between these two. To create this, iron makers of the Middle Ages adopted new technology: the finery forge.

In a finery forge, the carbon content of iron is controlled by bellows that force air over the molten pig iron, removing the carbon and phosphorous. Next, the iron is hammered (or wrought) to drive out the dross (silica bundled with other impurities). This process required lots of labor and wood, and although the bar iron it produced was a little harder and better than wrought iron, it was generally too inconsistent to be reliable blade material.

To make a better knife, we need consistent iron with an even more precise balance of carbon. In other words, we need steel.

Steel in the West

Generally speaking, steel is iron with less carbon than cast iron and more carbon than wrought iron. Exactly what percentage of carbon qualifies iron as steel depends on who you talk to. Steel is usually more pure than cast or wrought iron and is alloyed with other metals to help control ductility and strength, but for our purposes, we'll focus on the carbon.

Steel is the optimal blade material because it is strong and tensile, hard enough to hold an edge but soft enough to be tough and yield to a sharpening stone. It's the most popular metal in the world for culinary blades for good reason.

When steel first came about, adding carbon to wrought iron was nothing new. Humans had known how to do that since the Iron Age. But the means to do it precisely enough to make steel had to wait until the introduction of the crucible: a sealed clay or metal vessel that held molten iron or iron ore with a carbon source over an intense fire. South and Central Asia developed crucible furnaces in the medieval era, and in the postmedieval era, the West caught on, too. The steel that was made, called wootz, which originated in India, and Damascus as it was known when traded in the Near East, is some of the most famous in history. It was more versatile than previously smelted metals, which meant that a wider variety of tools could be made, and it had a remarkable mix of flexibility and strength. Oh, and the look of it is absolutely unmistakable.

The face of a wootz blade is a pool of wavy, swirling rivulets, layered in contrasting ribbons. The banded pattern emerges during forging if the timing and temperature of the heating and cooling phases are just right. Those snaking, rippling bands, according to researchers, are the alloys and impurities of the steel—including vanadium, cobalt, or manganese—that have glommed together into planes during the heating and cooling.

Because of its intrinsic flexibility, a wootz knife can be bowed over and bent far beyond the normal breaking point of other steels and spring back without damage. During the Crusades, as the story goes, Muslim fighters armed with Damascus swords cut the swords of Crusaders in half.

Wootz steel has a kind of mythic reputation that steelmakers have been unable to match since it died out in the seventeenth century. But in 2006, researchers dipped a seventeenth-century wootz sword into hydrochloric acid, dissolved the metal, and discovered a network of carbon nanotubes, each one about ten thousand times finer than a human hair. The theorized sources of those nanotubes come from the unusual carbon source—bamboo stalks, cassia wood, other organic matter—melted together with the iron ore. The organic matter contributes a tiny carbon structure of nanotubes that, if properly forged, align themselves around bands of cementite, a strong yet brittle intermetallic compound of iron and carbon. The tensility of the carbon balances the hardness of the cementite and together they make a hard but malleable steel. However, the exact formula and techniques for making wootz steel died with the last practitioner of the art, and wootz ultimately became a victim of the secrecy that surrounded its manufacture. Modern Damascus steel is a different thing, a pattern-welded or laminated blend of different metals.

Crucible steel advanced again in eighteenth-century Britain. Back then, the common steel in Britain was called blister steel, made by heating bars of wrought iron packed in powdered charcoal until they absorbed enough carbon to become steel. They earned the name blister steel (or shear steel) for the blistered surface they picked up during their long heating. The process was called cementation, and the blackish, bubbly steel it produced was unreliable and inconsistent because it absorbed carbon unevenly.

Benjamin Huntsman, an English watchmaker, was frustrated with the inconsistency of blister steel and the way it threw off the reliability of his watch springs, but around 1740, he discovered a way to refine it by adding a flux (chemical cleaning agent) that drew out the impurities that cementation couldn't. In his Sheffield steel plant, Huntsman used Swedish iron, which was less contaminated than British iron, and he heated his crucibles in a coke fire that reached 1,600°F/870°C. When the iron was melted and white hot, he added a little blister steel for carbon and then some flux to draw off the impurities. After a few hours at high heat, high-quality steel with uniform carbon distribution emerged from his furnaces. Huntsman's crucible steel was good and pure but difficult to make in quantity and expensive to buy. Other cutlers in Sheffield refused to use Huntsman's steel at first because it was much harder than what they were accustomed to, so he exported most of it to France, where it was made into top-notch cutlery that came back to market in Britain and outperformed the local knives.

The next lurch in progress came in 1856 when Sir Henry Bessemer, an English metallurgist, one-upped Huntsman's crucible with a new, hugely productive converter that could make steel so cheaply and on such a large scale that the steel could be used for everything from railways to pipelines. Bessemer's process shot blasts of air through molten pig iron, burning off excess carbon and raising the temperature of the iron. The technique produced massive amounts of steel with a fraction of the fuel and in a fraction of the time of previous methods. But there were downsides. Bessemer steel was high in contaminants (especially phosphorus, which makes the steel brittle) and only the highest quality (and most expensive) Welsh or Swedish iron could be used in the process. In 1876, Sidney Gilchrist Thomas improved the Bessemer process by adding limestone to the melt, drawing off the phosphorus, which formed as a slag on top of the molten steel and was easily removed.

By the mid-1860s, Bessemer had licensed the process to an American mechanical engineer who opened a steel mill in Troy, New York, that set in motion the growth of the United States as an important global producer of good-quality steel at a good price.

Bessemer's technique was eventually replaced, first by an open-hearth process, in which iron is melted in a large, shallow furnace, then diluted with wrought iron and oxidized to become steel, and then by the quicker, less costly oxygen furnace. But Bessemer still gets credit as America's godfather of steel.

Steel in Japan

The first signs of metalworking in Japan date from the very end of the late-Neolithic Jōmon culture. The subsequent era, the Yayoi period,

Bernal Cutlery's Steel

CARBON STEELS

Carbon steels contain less than 13 percent chromium. They're quicker to oxidize than stainless steels, and they react (by rusting) with foods that are salty, very acidic, or have a basic pH level. Carbon steels also require more maintenance than stainless steels and must be well dried after use to avoid rust. Carbon blades can take a finer edge than stainless steel and will hold it longer.

STAINLESS STEELS

A type of alloyed steel, stainless steels contain more than 13 percent chromium. No knife is truly rustproof, but chromium bonds with itself to form a film that reduces oxidation, making knife material that generally resists rusting. The downside is that chromium often worsens the performance of steel, and the stainless steels made thirty and forty years ago generally don't stand up to carbon steels in action. They are harder to sharpen, won't take as keen an edge, and they dull more quickly. These days, better stainless options are available, with a good edge, a long life, and easy sharpening. They vary greatly in their stain resistance, and all of them rust under the right conditions. Higher-quality stainless steels (higher in

set the stage for Japanese culture: rice growing developed, metalworking evolved, and kingdoms were established. In contrast to the West, there was no distinct Bronze Age or subsequent Iron Age in Japan. They happened at the same time, and the metals were used for different things: soft, ruddy, golden bronze was reserved for ceremonial, high-status items, and hard, democratic, everyday iron was the stuff of knives.

Whereas the West was able to rely on its own rich iron and other mineral deposits, Japan had only poor sources of iron ore. What it did have was something seemingly more awesome: the blackish iron sands in the riverbeds and rocks of the Chugoku region on the westernmost side of Honshu, Japan's largest island. *Satetsu* is its name.

A few types of *satetsu* exist, found only in Japan, New Zealand, and Canada, and they're all weathered from volcanic rock that cooled deep beneath the surface of the earth millions of years ago. To extract the iron from *satetsu,* in the sixth century, the Japanese developed a clay furnace, not unlike simpler furnaces in the West. Called a *tatara*, it had four walls, a floor of charcoal and ash, and openings on the sides to admit oxygen. It produced a steel called *tamahagane*, or "jewel steel," with an unusually perfect balance of carbon for making blades. The *tatara* and *tamahagane* drove the development of Japanese smithing for centuries.

The first definite evidence of smelting in Japan was discovered in the Onaru ruins, in Shōbara in the Shimane Prefecture, dating from the sixth century. During the Muromachi period (about 1336 to 1573), Japan was exporting a large number of swords to Ming-dynasty China, and as the demand for those swords grew, so did the size of the *tatara*. Although it

carbon and lower in chromium) tend to corrode more quickly if soaked or cleaned in a dishwasher. Despite what you may have read or heard, there is no such thing as a "dishwasher-safe" knife. There are only knives that degrade more slowly.

continued to change and evolve, the *tatara* remained at the center of Japan's steelmaking industry.

That is, until one dark night in 1853. On a quiet July evening, Commodore Matthew Perry and his fleet of gunboats chugged into Tokyo Bay to pressure Japan, which had closed its borders for some two hundred years, to resume trading with the United States. Soon after, Japan opened its doors and through them came the smelting technology of the West. By the turn of the twentieth century, Japanese steelmaking had undergone a transition from the *tatara* to productive Western furnaces that could supply weaponry for their own imperial army and navy.

In the early 1900s, the state-run Yawata Iron & Steel Co., Ltd., outfitted with German engineering, began producing steel on a large scale, and the country imported more and more iron, ore, and coke to feed the growth. By this point, steam engines had replaced waterwheels as primary sources of energy, and not long after, electrical power arrived, and with it the electric arc furnace, which was able to produce even more steel with less coal. The *tatara* faded from view, overshadowed by hulking Western-style furnaces, and Japan leaned more heavily than ever on foreign imports for raw material.

In 1950, the Korean War sparked a huge demand for inexpensive steel. To meet that need, industrial iron makers like Yawata reengineered their process to make steel more cheaply. Two decades later, Yawata merged with Fuji Iron & Steel Co., Ltd. to form Nippon Steel Corporation, which at its peak in the 1970s was making 47 million tons of steel per year, passing the United States Steel Corporation as the world's largest steelmaker.

These days, the *tatara* still exists, but given the enormous amount of skilled labor it requires, its low yield, and the hazards involved in working one—*tatara* masters eventually lose their vision from staring into the blazing flame—it's not all that surprising that the production of *tatara tamahagane* is highly limited. Nowadays, high-quality cutlery steel is made by manufacturers who can approximate some of the qualities of *tamahagane* by performing finely tuned modern extraction processes on *satetsu*. The most notable of these is Hitachi Metals, which makes much of the steel you'll find in the knives we sell at Bernal Cutlery.

Today, tamahagane's offspring, industrially refined sand-iron steels, live on in a few favorite Japanese cutlery steels: aogomi and shirogami.

Chapter 2

Europe

A century ago, European knives were front and center in the Western culinary world, but they've been overshadowed these days by our new five-hundred-watt obsession with the Japanese knives that retained lightness and agility lost to modern, mass-produced European knives. But to talk about the mass-produced European knife as emblematic of a long tradition of handcrafting is way too simplistic. Luckily, there are a few holdouts that have kept the traditions of European handcrafted cutlery alive.

If Western knives have a trademark, it would be their utility. They can jump from a lamb leg to an onion as if it's nothing, while a Japanese vegetable knife can't handle sinew and bone. European knife history has none of the mystical cachet that samurai swords bring to the Japanese story—a narrative knife sellers like to lean on too much—but the European blacksmiths who banged out swords in the Middle Ages were actually creating some fairly complex tools. They worked with a carefully balanced cocktail of metals layered together in a process called forge laminating that was largely abandoned as steel and technology improved. These days, Japan is better known for its forge laminating than Europe.

Bit by bit, when industrialization blazed through Europe and brought machines and mass production, European knives traded their finesse for market share. Losing handwork stripped European knives of their subtle curves and lightness; cutting costs and adopting automation increased production volume but at the cost of Europe's reputation for fine cutlery. By the 1990s, Europe had all but abdicated its cutlery throne, and when Japanese knives came ashore, it was theirs for the taking.

The story of European knives develops differently from the story of knives in Japan, of course, but what I find interesting is where their stories link up. When they do, we get knives with the best sensibilities of both Japanese and European styles: utility, precision, and lightness. They are knives that want to do a lot and do it well—knives that want to work, and work with intention.

When I decided to write this book, I didn't count on bringing Europe into the heart of it. Sure, the shop carries some excellent knives from France and Germany, but the Japanese knives were always its focus. The thing is, whenever I followed our most popular Japanese knives to their root, I landed in Europe. And when I went to visit the knife makers in Solingen and Thiers, I left with a much deeper appreciation for Western knives and for the Japanese-made styles they inspired. In the end, I realized that it's impossible to tell their stories separately. The more we flesh out each of their histories, the less distance there is between them. Before we dig into those stories, let's look at some common Western knives.

Western Knives Primer

CHEF KNIFE

A chef knife is widest at its heel, with a blade length of 6 to 14 in/15 to 35.5 cm. The most common sizes today are an 8-in/20-cm blade for home kitchens and a 10-in/25-cm blade for professional kitchens. Chef knives are most often used with a forward slicing, or "rocking," motion, in which the tip keeps contact with the board, or less often with a pulling motion that moves the knife backward across the food, working from heel to tip. When used with a rocking push cut (see page 152), the chef knife is a very powerful cutting tool. It requires a fraction of the motion that another knife, like a slicer, needs to do the same work. This is the real strength of the chef knife and how it is able to work quickly and effectively. It works well on vegetables, but it's also a good tool for butchered meat and other proteins like fish.

FRENCH CHEF KNIFE

The French chef knife has a more triangular footprint than the curved belly of its German equivalent—and a distinctive handle, with a single piece of wood (often ebony, rosewood, or sometimes European walnut) with a rat-tail (stick) tang that runs all the way through the handle's length and a metal ferrule ring.

The rise of the hotel restaurant and the growing popularity of the French cuisine it served encouraged sales of the French chef knife in the twentieth century. In the period following World War II, the appeal of French food continued to grow, propelling French knives into more and more kitchens, even as their actual production numbers were dwarfed by German knife maker Solingen.

GERMAN CHEF KNIFE

The German chef knife is generally thought to be heavier than the French version, though this is only true (well, mostly true) if we're talking about the widespread modern versions. When Solingen was the global center of knife making in the late nineteenth century, it absorbed influences from all over the world. Distinct regional styles from France, England, and America quickly populated the outward-looking Solingen knife maker's catalogs.

One characteristic common to German knives is a slightly more rounded edge and a slightly squarer bolster, the junction between the handle and the flat of the blade, than the French, which has evolved to carry a rounder bolster and ferrule, a ring of metal where the handle and blade meet. The chef knives made in Solingen became some of the most widely used chef knives in both home and professional kitchens starting in the 1970s, and they dominated the scene in the 1980s and 1990s, by which time they had grown heavier with the virtual disappearance of handwork.

ENGLISH CHEF KNIFE

The English chef knife is the hen's teeth of vintage chef knives. They do exist, but they're rare. At first glance, an English knife looks like a French one, but the flatness of the blade's heel, the oval ferrule and bolster, and the smaller bird's-head pommel, the counterweight at the end of a knife's handle, give them away. There are not many of them because Sheffield, the old epicenter of England's knife production, produced few. Many of the distinct cuisines of England were lost in the mass migrations to large cities during the Industrial Revolution. As more people were shunted off into factories and both parents in a family worked, the need for a chef knife at home vanished, while a can opener rose in its place.

The footprint of the English chef knife suggests the influence of the large roast-carving knives that owe their existence to a fifteenth-century English culinary invention, the roast beef. Back then, England was rich in wood fuel and cattle, and roasting joints of beef by the radiant heat of a big fire was a common pastime across the country. The breeding of cattle for meat began in England, and many of the breeds today that are synonymous with beef are old English ones. The carving knives evolved into wide slicers with a long edge that ran parallel to the spine, ending in a clipped point. Oddly, the trademarks of an English chef knife—the oval ferrule and bolster, the pommel—can be found in American and some German knives from the late nineteenth and early twentieth century, too, proving just how promiscuous cutlery producers were back then.

SLICER

Here is another design as old as dirt. Slicers came about as people began making knives out of material other than stone, which is brittle when fashioned into a long blade. They are narrow knives designed especially for proteins, typically used with a pulling stroke from heel to tip. Slicers are similar in length to chef knives, between 6 and 14 in/15 and 35.5 cm, though they typically run on the longer side because they are used with a pulling, rather than a pushing, motion. The narrow blade offers less friction than a wider blade, which is a distinct advantage when cutting animal protein.

Slicers come in a broad range of thicknesses and blade shapes. Narrower blades drag less and make curved cuts more easily. Wider blades will drag but cut straighter, which is good for slicing thinner cuts of meat. Interestingly, there seems to be far more variations of the slicer than the chef knife. The medieval roots of the slicer can be traced to the *tranchelard*, the long, narrow-bladed, pointed knife that regularly accompanied a dish of trenchers, a large, flat wheel of bread topped with meat and sauce. Flexible slicers are represented today by fillet knives and ham knives designed for working around the bones of bone-in fish and ham.

CARVING SETS

Carving sets are interesting, both as a player in cutlery history and as an emblem of change. They were centerpieces in the old tradition of Sunday-night dinners, but those have largely faded into the rearview mirror of life in the West, certainly in Europe and the United States.

Nowadays, these sets are most often regarded as heirlooms. In families that still use them, they might come out of the credenza only two or three times a year if used very heavily; otherwise, once a year is average. Carving sets evolved differently in different places, though they were generally displays of hospitality and, sometimes, of the host's wealth. Many people bought the finest they could afford.

ENGLISH AND AMERICAN CARVING SETS

By the late twentieth century, a standard carving set included a knife, a fork, and a sharpening steel. Deer antler was a popular handle material, as were exotic wood, silver, mother-of-pearl, and bone. Although the knives in many earlier English and American sets had a straight blade, by the end of the twentieth century, a curved, pointed blade had become the standard. Many upscale sets featured knives and forks of different sizes for larger or smaller cuts of meat and for birds of all sizes. The silk-lined, leather-bound boxes of the nineteenth century could hold seven or more pieces of cutlery, while the boxes of more modest sets held just three.

FRENCH CARVING SET

France is home to a unique carving set designed specifically for lamb: a 7- to 8-in/18- to 20-cm knife, a three-pronged fork, and a pronged clamp with a screw that tightens its grip on the bone of the *girot*. To the uninitiated, the bone holder might look more like some genteel version of a medieval thumbscrew torture device. The French sets may also have a curved- or straight-bladed knife with a silver or white metal ferruled handle made of walnut, ebony, ivory, horn, silver, or composite materials like celluloid or Bakelite.

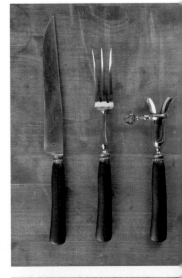

SAN FRANCISCO CARVING SET

Nineteenth-century San Francisco might be famous for the gold rush of 1849, but it was the massive silver deposits of the Comstock Lode that built the city in the ensuing decades. In those prosperous years, the First Transcontinental Railroad had not yet been built, and all imports had to make the expensive trip around Cape Horn to San Francisco. That pushed San Francisco to develop its local industries, among them cutlery. The city's cutlers came up with their own style of carving set: an offset narrow blade with a saber grip and a handle of walrus ivory or stag horn, with fork and sharpening steels fit with four silver-tipped "antennae." In time, their unique design became popular outside of San Francisco, and large East Coast manufacturers were soon imitating them, albeit at lower quality.

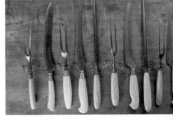

UTILITY KNIVES

OFFICE, UTILITY, PARING KNIFE

Pointed, agile, and just 5 to 6 in/12 to 15 cm long, the utility knife is possibly the most ubiquitous class of kitchen knife, as its ambiguous name indicates. The term *utility* refers to a general size of blade that has been with us since the small, flaked stone tools of the Stone Age, with later ancestors represented in knives the world over, including those living on the hips of medieval Europeans and persisting into the kitchens of twentieth-century Paris as the "office knife." Castles and manorial houses in late-medieval Europe reserved a special room, the office, for preparing fish or game, and the utility knife that was used there (and took its name from the room) was narrow, with a pointed tip for working around small animal bones. French office knives survived into the kitchens of the famous early nineteenth-century pastry chef Antonin Carême, and later, of Auguste Escoffier, usually as the counterpart to the chef knife—a holy culinary pairing that many chefs still rely on today. The office knife eventually evolved into the Western utility knife, and from there into the Japanese petty knife, all of which are best-selling knives today.

Paring knives are a smaller derivative of the office knife, used to peel things in hand, with a blade 2½ to 3½ in/6 to 9 cm long. Any longer and

the tip would be too far away, forcing the hand to slide up the blade as the knife is used.

BUTCHERY KNIVES

Knives for butchery vary greatly by location and the task at hand. Smaller knives are better for seam butchery (similar to older styles of butchery), or taking an animal apart along the creases of its musculature and the weak parts of its joints. Butchering requires a firm grip, and many butchery knives developed a larger handle.

The *boucher*, an old style of larger butchery knife with a wide blade that ends in an upswept tip, is used for both marking where to start large cuts and for cutting steaks and chops. A possible progenitor of the chef knife, the *boucher* took different regional styles in northern and southern Europe but has survived in France and parts of southern Europe for centuries with very little change. It has a blade 6 to 14 in/15 to 35.5 cm long and a one-piece wooden handle riveted to a tang that runs three-quarters of its length. The northern European–style "Alsatian" *boucher* has given way to the scimitar (also spelled cimeter) and bullnose butcher knife but has persisted as a farmhouse butchery knife for centuries. Its defining feature is a full tang with riveted handle scales, as opposed to the blocky solid wood (or now composite) handle with carved grip. The distinctive feature of the *boucher* is the straight spine and an edge that curves up somewhat dramatically to form the point. The round edge is ideal for drawing a slice of meat because the contact between blade and meat is focused on a small area, concentrating the pressure of the blade. The regular European *boucher* and its Alsatian sister are always found with simple, economical handles. They're never made with fancy materials or with ornate grinding or etchings. These were tools for butchers or farmers who used them in the era before band-saw butchery and for farmers who on occasion broke down large animals.

BREAD KNIVES

Bread knives are wallflowers, with unassuming looks but a rich past. Today, the serrated bread knife is more or less sacrosanct, though it did not achieve that status until the twentieth century. In times past, for better or worse, bread formed the bulk of the working-class diet. In Germany, the bread served to farmworkers in the fields was cut in hand off a flat, round loaf that was pressed to the chest in a headlock (or shall we say, breadlock), with the knife held backward so the blade faced the arm. The loaf was cut by pulling the blade toward the body, and the bread knife from Hamburg developed a footprint to suit: a short, wide blade about 7 in/17 cm long, with a curved spine and a handle with a reverse curve. Europe and Great Britain had similar versions in the nineteenth century.

Serrated blades didn't begin appearing until the late nineteenth and early twentieth centuries. Early versions featured a wavy pattern, a series of slits that made a sawtooth edge, or tiny, pointed saw-like teeth that alternated directions. Some mimicked a slicer's curved tip, while others had a wider heel, like a chef knife. Oddly enough, it was the influence of the German

Hamburg-style blade that inspired most modern bread knives, even though they are used on a board, not in a "breadlock."

THE NETHERLANDS BOSCHER

The Netherlands *boscher* is hard to categorize, but it will live here in the bread knife section because it has been used heavily in that capacity and it shares some similarities with the other northern European knives, like the Hamburg knife. It's one of the only knives for which the DNA has hardly changed since the Middle Ages. The blade is narrower at the heel than at the tip, where the spine curves down to form a point, like a very slim-waisted sheep's-foot blade.

The Evolution of the Chef, Abridged

We'll start the story of the chef knife after the Bronze Age and medieval times in Act Two: The First Restaurant. The first business to call itself a "restaurant" was a French establishment run by a M. Boulanger, who opened a shop in 1765 to sell rich, strongly seasoned "restoring" broths. Although they had been popular as medicinal tonics for hundreds of years before this point, they hadn't been sold in shops like this. The broth model caught on quickly in Paris, but selling most kinds of food beyond broth was still strictly regulated by the separate guilds of the day: butchers, bakers, pâtissiers, and caterers.

At the time, dining options for the common people were limited, and private chefs were at the disposal of the wealthy. There were street vendors who sold snacks to eat on the spot, inns that served travelers a modest set menu, and caterers who arranged meals to order, although they were pre-planned and never executed on the fly. Taverns where alcohol was sold had meals from local caterers or inns' kitchens on hand, but the offerings were few and always precooked.

The French Revolution brought a big shift in the way dining happened. For the first time, restaurants began serving from an à la carte menu of separate items with prices. It was a giant change in both the concept of where and how eating happened in public and how food was prepared for the public.

After the French Revolution, cooks from the kitchens of estates of deposed aristocrats began flooding into Paris and other cities, energizing the growth of à la carte restaurants. The enthusiasm for English styles (especially the tavern) dovetailed nicely with the revolutionary spirit of the day:

common people were now elevated to what had been until recently the provenance of only the rich and powerful. Now a chef-made dinner was for everyone.

But à la carte cooking required a new conceptual tool kit, and a new knife kit to go with it. Paintings of cooks from the turn of the nineteenth century feature two knives at the ready in their waistbands, a chef knife and a smaller office knife. Maybe unsurprisingly, these remain the most common knives in the knife roll of a Western chef today.

The chef knife proliferated at this point because it could cross the boundaries between produce and protein, perfect for the needs of à la carte service where proteins and vegetables were prepared, cooked, and assembled all at once. And as more and more à la carte restaurants opened, orders for chef knives poured into knife-making towns like Thiers and Solingen. But no one did more to expand that need than Auguste Escoffier.

Escoffier modernized and simplified French cuisine. He emerged in post-Revolution Paris at a time when the old style, as exemplified by the extravagant culinary stylings popularized by pâtissier Antonin Carême earlier that century, was still in vogue. Escoffier didn't see the point of it. He turned to Paris and told the people to *faites simple*—"make it simple."

Escoffier defined a new era of French cooking and laid the groundwork for the professional kitchen as we know it today. He wrote down recipes and codified measurements and steps. He divided his cooks into a hierarchy of positions, a brigade system of specialized roles: the *chef de cuisine*, the sous-chef, the *saucier*. And he demanded that everyone in that system wear a hat and act dignified. He cracked apart the caricature of the drunken, screaming French chef and replaced it with a calm, famously unflappable standard. If he ever did feel a tantrum rising, he left the room.

The brigade system allowed chefs to delegate the components of a single dish—protein, vegetables, sauce—to different cooks, allowing for a smoother, quicker assembly than one person cooking from start to finish could manage.

Escoffier's brand of restaurant caught fire. He was a brilliant and progressive marketer, and in a time when it was scandalous to find women dining out, he made the new Savoy in London a respectable and welcoming place to do that. He named his dishes after famous guests to keep his restaurant in the public mind and eye, and as more restaurants opened in the brigade model, France entered its knife-making prime, most notably supported by the verdant, riverside knife-making village of Thiers.

Thiers

A small, charming medieval town in the Auvergne, Thiers sits along the steep banks of the Durolle River, which rushes down from the Bois Noirs forest. The Durolle was the lifeblood that powered the forges and grinding houses here, fueling a network of artisans scattered throughout the hills. The output of Thiers has always been a fraction of other knife-making towns, such as Sheffield in England or Solingen in Germany, but even today, two-thirds of all the knives made in France bear the Thiers stamp. And Thiers still relies on a network of independent artisans.

Painted in red and mustard yellow, Thiers stands in the hilly terrain above the small plains below the Puy de Dôme, a volcanic mountain in the Châine des Puy range of central France. Here and there, box stores lurk in the valley at the foot of the town, as if they're about to lay siege to the stone and wooden houses on higher ground. The workshops remain scattered in the hills above the river, some of them working and others quiet, dusty monuments to an old industry.

Wedged in these hills is K Sabatier, a legendary knife-making company that has been run by the Bournilhas family for eight generations. Many small knife makers here attach "Sabatier" to their name—it's an untrademarked sign that's meant to imply a high-quality, hand-forged knife—but K Sabatier is the longest running bearer of the Sabatier name, headquartered in a modest, two-story white stucco place, with the workshop time forgot around the back.

These days, Philippe Bournilhas, president of K Sabatier, runs the operation, though he prefers a more modest title: the "eighth generation" of his family to helm the business. His father, Alain, takes the place of the courier, and Christine, Philippe's sister, runs the family retail shop in nearby Clermont-Ferrand. Alain is supposed to be retired, but he still makes the rounds in his van to the different workshops that help to piece together K Sabatier's knives. The craftsmen around Thiers get to choose with whom they work, and the number of longstanding and loyal contacts the Bournilhas family has is a remarkable testimony to their character. And after fifty years of delivering knives between craftsmen, Alain knows more than anyone about the history of knife production in Thiers.

I've come to K Sabatier for the sales ledgers, and Philippe and Alain know it. They've searched through the archives to dig up the records that cover the period from the 1890s to World War I, when the hotel restaurant was at its height and Escoffier was running the show, in hopes of helping me sniff out the chef knife.

I found orders for chef knives—for knives to Paris, London, New York, and even San Francisco. But the most important thing I found illuminated more about Bernal Cutlery's relationship to the past than any epiphanies about the chef knife could. Among those countless rows of tiny, florid cursive were patterns I recognized from our own books: large orders for office knives under 6 in/15 cm and lots of chef knives that were 9 or 10 in/23 or 25 cm. The most popular knives in Bernal Cutlery today are just those: the 240-mm/9½-in *gyuto* and the 150-mm/6-in petty knife, Japanese renditions of the chef and office knife. The needs of the kitchen, it seems, do not know time.

While we pored over the dusty books, Alain and Philippe unearthed a few half-finished, misshapen, and dirty knives from the next couple of decades. These lumpy relics were time capsules that captured the evolution of Thiers knife making, since the nineteenth century. The key to understanding them is in the bolster, the junction between the handle and the blade.

ACCORD DU 15 JUIN SIGNE A LA MAIRIE DE THIERS

Entre les délégués ouvriers d'une part,
M.M. CROSSON, Secrétaire général de la Bourse du Travail,
GUILLAUMOT, Secrétaire du Syndicat unique,
GASCHON et BECHON, Délégués du Syndicat Unique,
DUMOUSSET et LOMBARDY, mandatés par le Syndicat des Emou

et d'autre part,

Les Délégués de la Chambre Syndicale de la Coutellerie

French bolsters changed at a different pace than German ones because the drop forge came into prominence later in France than in Germany. Before World War II, most French knives were forged and ground by hand, their bolsters thin and narrow. After the war, when the drop forge came on the scene, the bolster and ferrule ring were joined into one solid drop-forged piece that became the hallmark of the postwar French knife. Up until 1990, trip hammers still tapered tangs and thinned blades. Smaller knives were first made with a drop forge from start to finish, and the very last blades to see the trip hammer before drop forging took over completely were the giant 12- and 14-in chef knives.

Down the valley, away from the twisting medieval streets of Thiers, is a drop forge. The French and German forges—which you'll meet shortly— each serve their town in a similar way, as the anchors of the smithing network. The forgings they make trot through half a dozen workshops in the hills before a knife is finished. Neither of them attempts to compete with the web of craftspeople that surrounds it. They stick to forging, as they've always done.

The History of Knives in Thiers

If we trust the old grinding stones in the walls of Saint-Genès church, cutlery production in Thiers started in the thirteenth century. Back then, iron and steel and the sandstone for grinding wheels were brought into the Durolle Valley from Bourgogne, Dauphiné, and Nivernais. Thiers had all the trappings of a knife-making town: water power, timber, and plenty of people to learn the skills. In a short time, small ateliers rose up along the river's edge, and workshops that needed no water power found clearings in the surrounding mountains. Today, the town of Thiers remains surrounded by small homesteads and workshops strung together by narrow roads and centuries-old footpaths.

Laborers in Thiers have always been quick to organize, starting in the sixteenth century when they coalesced to form the Jurande, a guild to oversee the quality and production of knives. The Jurande established trademarks and codified the steps to becoming a master cutler: three years in a workshop, five years as an apprentice, and however long it took to forge a knife that passed muster with a senior cutler. Master cutlers had to be born locally and at least twenty-four years old. By the late 1500s, knife makers were selling their wares from storefronts in Thiers, some of which still stand today.

The Jurande also established trademarks to brand and market knives from Thiers: images and numbers that a society with a low literacy rate could recognize. Knives were stamped with signs of the master cutler who oversaw and vouched for the skill of the maker. Trademarks verified the quality and spread the fine reputation of Thiers as an exceptional knife producer. Even in the 1500s, marketing was as vital to business as it is today, and the Jurande knew it.

Trademarks were stored in a locked table, and new designs could be submitted only once a year, on May 1. On that day, the five members of the Jurande who each held a key to the table would meet and enter the new trademarks into a lead tablet held in the table. All knife makers were required to mark the knives they produced with their trademark and include this trademark on their shop sign, along with pictures of the items they created, such as knives, razors, scissors, and so on.

But life as a cutler in Thiers wasn't all cozy knife shops along the quaint medieval streets of the Durolle Valley. Grinders worked lying down over their wheels, pressing blades on their grindstones from above to utilize the full force and strength of their arms and body weight. Workplaces were rough and cold, and dogs were trained to lie on the legs of grinders to warm them and keep their blood moving in the winter months.

Despite these harsh conditions, the grinders worked quickly and effectively and were regarded as the top skilled laborers in Thiers. That meant they were able to organize to buy their grindstones, stipulating prices and enforcing quality-control standards. They were also adamantly anti-Catholic and chose Monday, instead of Sunday, as their day off to make a show of not going to church. The life of a grinder was physically taxing, strained the neck and shoulders, stressed the ribs, and it put the practitioners at risk of catastrophic injury. Sandstone grinding stones could explode when spinning at high speeds because of the small cracks they acquired during transport from the quarry. Even in the best of circumstances, stone powder from the spray of a fast-moving grindstone left many with silicosis, a lung disease, landing them on their deathbeds by their sixties.

Grinders suffered, but they also enjoyed a long tradition of autonomy and independence, and as contractors, they chose their own clients. The

economic realities and revolutionary attitude of the late nineteenth century eventually led to widespread labor organizing, and the grinders of Thiers gained a reputation as especially strident union organizers.

Even though the output of Thiers has always been small compared to major knife-making towns, the knives it did make had a tremendous impact on the proliferation of French knife styles around the world and inspired the manufacture of a huge number of French-style knives in an epicenter of knife production: Solingen, Germany. In its prime, at the turn of the twentieth century, Thiers was a prosperous town that fed not only the skyrocketing demand for knives in professional kitchens like that of Escoffier but also for scissors, razors, and the huge variety of pocketknives for which it remains famous today.

The Rise of the Chef Knife in Thiers

The end of World War I brought France victory, but at a steep price. Some 10 percent of the male population had died in the war, and large stretches of France had been scorched black. But robust growth in the 1920s carried France through the crisis of 1929, Paris became the cultural capital of the West, and with the expansion of French food, orders for French knives poured into Thiers. This is where Thiers sets itself apart: Even after the war and the surge of industrial growth, the town held proudly to its heritage of handcrafting goods. The craftspeople were still in peak form, and the knives they produced with old production styles during this period—forged and ground by hand, using waterpower from the Durolle as well as newly abundant electrical power—were, and continue to be, absolutely superb. They felt great in the hand and cut smoothly thanks to the fine grinding. Many of these were chef knives, bound for the cooks of the now canonized brigade system.

They were light and nimble, forged and thinly ground. They had narrow bolsters with a rat-tail tang (the narrow extension of the blade through the full length of the handle), set into a single piece of tropical hardwood with a brazed steel or nickel silver ferrule collar, unlike the riveted handle and rounded integral bolster we see today.

In Thiers, this style was referred to as the *cuisine massive*, to play up the strength of the thin knife, and possibly to deflect attention from the rival town of Nogent, a cutlery center in eastern France.

Thiers prospered in the exploding demand for its knives and even remained hardy after the economic devastation of the post–World War II era in France. While much of the rest of postwar Europe was forced to reinvent itself, faced with intense poverty and collapsed infrastructure, Thiers pulled itself up by its bootstraps and got to work, thanks to the resilience of the small atelier networks.

Thanks to Julia Child and Jacques Pépin, the United States saw a surge in enthusiasm for home cooking in the 1960s and 1970s, and with it, a new taste for French carbon-steel knives. The excitement lasted until 1976, when French law banned carbon steel in professional kitchens in favor of easier-to-sanitize stainless-steel knives. That government action halted carbon-steel production in Thiers, and the lower-quality stainless knives that replaced it started to chip away at France's gleaming reputation for quality, high-performing knives.

With carbon steel fading from view, cooks reached for German, Japanese, and cheaper American knives, which forced the cutlers of Thiers to diversify their production. They began forging bicycle components and industrial goods, and they kept their small atelier system intact.

In the 1990s, just as Thiers was getting its sea legs again, a second wave of exports came onto the market, this time from China. Big-box stores moved into town, starving out the small businesses that were once supported by the local workers. But, as always, Thiers has proved resilient. Those grinders and fitters and finishers are still toiling quietly in their homes and workshops at the ends of roads no wider than the bumper of a tiny Renault, grinding pocketknives and riveting wood handles onto slicers. These days, there is neither a formal apprenticeship program nor a guild, but knife makers have pooled resources to create a thriving town brand and knife style: Le Thiers.

And while the culinary knives of Thiers have embraced some new designs to remain current, they've also stayed true to their carbon-steel roots. That loyalty has gained a new following in the United States that Bernal Cutlery is proud to be a part of. Thiers is the land of Vercingetorix, the great Gallic leader who repelled the invading Julius Caesar, and as much as other centers of cutlery might dwarf it, the city has never fallen without a fight.

Solingen

As far as European knife history goes, it would be next to impossible to overlook Solingen, in the hilly Bergisches area of western Germany. The local mountain soil here is poor for farming, but the hills are rich in iron

ore, the forests are thick enough for making charcoal, and the Wupper River flowing through the valley provides a constant source of power. It's a lush and fertile landscape for smithing.

Solingen was once the largest producer of knives in the world and still produces many of them, but the town was known for weaponry—swords and the like—long before it made its name in culinary knives.

For the bulk of Solingen's history, the Wupper River was the engine that powered the city's numerous small riverside workshops, or *Kotten*. The blades forged in the *Kotten* were carried uphill first to the grinders, and then to the fitters, who installed all the parts to make a finished product and worked from their homes, which were clustered together uphill from the *Kotten*. The workers themselves were called *Heimarbeiter*, or "home workers."

The nearby Rhine River led to centers of commerce, and Solingen made its name with the quality of its blades, especially the swords, which were sought by buyers all over Europe. Smiths, grinders, and fitters specialized in specific types of bladed instruments: kitchen knives, pocketknives, scissors, swords, and so on. Workers were self-employed, and if they didn't own their own workshop, they rented space or bought space from a larger shop (some grinding *Kotten* had up to a hundred grinding stations). Traditionally, workers were paid by the piece, and this continued through industrialization and unionization.

As business boomed, small clusters of homes called *Hofschaften* grew up around the Wupper Valley. These belonged to *Heimarbeiter* who didn't require waterpower for their work, so they built their houses and small workshop uphill. As the number of riverside *Kotten* grew, so did the *Hofschaften*, and gradually more small workshops and homes popped up along the roads connecting different settlements. Eventually, Solingen grew into several larger centers: Solingen-Dorp, Gäfrath, Merscheid-Ohligs, and Höhscheid.

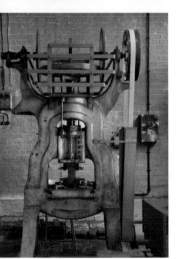

In 1867, the Solingen area was connected to the Cologne railway, which made the transport of large quantities of coal and other supplies to the city possible. With coal, steam power began to replace waterpower, and soon larger factories were born on the high ground of the small mountain ridges. In the 1880s, newly developed steam-powered drop forges and die stamp hammers moved in and production exploded. Before drop hammers, two smiths provided enough rough blades to keep up with three grinders, but a drop forge banged out enough work for five grinders to finish. The drop forge is still used today to shape hot steel with a mechanical hammer, pounding it into a mold cut into an anvil. The rough forging is then taken to a mechanical punch to remove excess metal, and from there it is hardened, ground, and finished.

The independent specialized workers in Solingen today mirror the old *Heimarbeiter* system. Some of the forging is done mechanically, but the final finishing is still largely done off-site. Holding on to this preindustrial

legacy is what helped Solingen maintain its place as a center of skilled labor and of innovation. The growth of mechanical forging created a bottleneck of available grinders, and in the 1920s, some moves were made toward automation. But those machines still needed skilled operators. Only after World War I, when faced with a shortage of workers, was mechanization more enthusiastically utilized, and even then, skilled labor remained at the heart of Solingen production until later in the twentieth century.

In an environment of independent workers, it is both difficult to keep a trade secret secure and easy to spread information and new techniques. Solingen was a hard place to keep a secret but became a great place to produce things because new skills and improved techniques spread like wildfire. Just before World War I, Solingen was exporting 75 percent of its production and supplying nearly 60 percent of the global market.

Rudolph Broch

The old *Hofschaften* and *Kotten* scattered along the river and throughout the hills of Solingen still exist. Some of them have been abandoned, but not all. Inside one of them is Rudolph Broch, a knifesmith and guardian of the old Solingen way—one of the few who remains.

Rudolph Broch's workshop is a cluster of redbrick buildings filled with old equipment, most of it unused, as the workers who know how to operate it are all gone. The machine he works with is called a *Fleishmeider*, a mechanical hammer set at the end of a large wooden arm that drops as a one-cog wheel turns. The hammer and anvil are oriented to draw the blade along its width, rather than its length, which aligns the latticework of the carbon along the edge and makes the steel tough and durable, ideal for the small high-carbon blades of pocketknives. With a hard high-carbon steel, this kind of hand forging dramatically elevates a knife.

Rudolph Broch, with his bright blue eyes and thick workman's knuckles, is one of the last smiths in Solingen to operate a flat forge, and he might not be here were it not for Giselheid Herder-Scholz. Giselheid is the managing director of the Robert Herder company, which has been in her family since 1872. But the Herder family's legacy as cutlers is far older than that; in the Solingen dialect, their surname means "hardener," referring to the job of tempering the steel after forging. Giselheid has been instrumental in preserving the legacy of old Solingen knife-making techniques, and she revived Rudolph's business when it was in peril of disappearing. When she found him, she commissioned an ongoing order for a short, high-carbon chef knife that was exceptionally durable, thanks to the unique flattening pattern of the *Fleishmeider*. Although he's nearing the end of his career, Rudolph Broch is still very much at home in front of the forge, feeding small billets of steel under the well-placed strikes of the *Fleishmeider*.

Rohrig Forge

A short drive through the rolling green hills, away from Rudolph Broch's dusty workshop in a shingled neighborhood in Solingen, is the Rohrig Forge. Rohrig is the portrait of large-scale forging, but it operates much like the early riverside forges did in past centuries. The knives start here, but then are dispersed to artisans who complete the process.

Rohrig is a clanking, pounding, industrial machine. Housed in a large, new building in the Merscheid neighborhood, it would look right at home in an American office park. Sections of steel are sheared off long, square billets, and half a dozen drop forges pound away at the same time. Workers move glowing hot steel from a furnace to the die of a forge. The hammer deals a blow then bounces back up, and the die-shaped workpiece is dropped into a crate, where it darkens and cools from red to black.

In the next room, shearing machines stamp unhardened forgings, which are quite malleable at this phase. A hulking CNC (computer-controlled) machine sits in the corner, waiting to turn out new dies after a few thousand punches register the old ones defunct. Pulling more of the steps into their own house, like handle fitting or grinding, would be a profitable move. But when I ask them why they don't, they frown and say, "Why would we compete with our customers?"

The Robert Herder Company

Back in the Ohligs neighborhood, Robert Herder Windmühlenmesser keeps their process in-house from start to finish, except for forging. It's still the largest operation in Solingen that keeps its focus on old-style handwork.

I remember the first Robert Herder Windmühlenmesser knife that I ever picked up. I plucked it from a pile of knives at a flea market, and I was amazed by its lightness because I'd always thought (as so many people do) that modern German knives were heavy, industrial things (of course, many of them are). I'd find out later that lightness was in the DNA of Robert Herder knives, thanks to the handwork of *Heimarbeiter* like Rudolph Broch.

Centralizing the process this way, keeping all the steps except smelting and forging in the same place, is an anomaly in Solingen, but it has its upsides, too. By housing the whole process in one place, secrets are better protected, and old techniques can be salvaged from skilled workers who are nearing the end of their careers, and steadily disappearing. As an independent contractor, there's an advantage to playing your cards close and keeping your recipes under lock and key. But when that handle fitter or grinder passes away, and his workshop closes, those secrets dissolve in the ether. Robert Herder Windmühlenmesser keeps those secrets alive through an apprenticeship system, regularly recruiting students to train with old masters, absorbing their secrets and skills for posterity.

Between 50 and 100 percent of the finishing process here, after forging and smelting, is done by hand. The knives are made thin, as most high-quality knives were in Solingen before the 1970s and 1980s, when faster methods like automated grinders came along. Solingen even has its own term for a thin edge: *nagel-gehundt*, or "nail-going." In other words, it is so thin you can bend it, if only temporarily, by running a fingernail across it.

If Robert Herder Windmühlenmesser knives are uniquely thin for European knives, it's partly because they're only *mostly* European. Giselheid Herder-Schulz has been traveling to Japan since the 1990s to collaborate and study with Japanese smiths and sharpeners, and over time, she's been infusing the Robert Herder practical manifesto with Japanese values—namely, the cutting feel of a knife. Japan never lost its focus on the superior feeling of a thin, convex, smooth-cutting knife, the kind that slides through food like butter and carries the reverberations of texture into the palm with a clarity that machine-made blades just can't match. When Solingen embraced industrialization, it lost that focus. But thanks to Giselheid, it's coming back.

Robert Herder Windmühlenmesser captures one hundred years of technology in the process between a chunky, charred blank and a fine, nail-going finish, which is at the heart of what makes this place so unusual in Solingen. Old-style handwork that was discarded in the wake of World War II is alive here. Some of the machines are at least eighty years old, but they look more burnished than they do dirty.

The process starts with blades that are either stamped from enormous rolls of sheet steel or forged somewhere else and delivered. After that, the blades roll through the heat-treating furnace, are cooled in a vat of oil, and then are brought to a separate annealing oven to lock in a fine-grain structure and hardness. The first rough shaping happens on a new CNC machine that carves out the proper shape and distal taper from heel to tip. Next, depending on the knife, the blade may head to a 1920s Kessels grinder. The wobbly grinder, as I like to call it, developed as a safer alternative to hand grinding on large stones. It holds the blade in an arm that wobbles against the flat side of a large whetstone, enabling the operator to set the convexity of the face. I like the charm of small machines, and few large machines make me feel jealous—the CNC machine interface gives me a headache—but the Kessels grinder, the love child of a Volkswagen Beetle and a grinding wheel draped in gritty towels, stole my heart. It may be one of the last in operation and most likely the last in regular use.

After the initial shaping on the wobbly machine of my dreams, the blades head to a tall, narrow wheel made of linden wood and walrus leather—thick enough to hold up but soft enough to give a little during use. The blades are given a fine glaze of abrasive, and then a finer blue glaze that gives them a bluish reflection. This signature Solingen "dry fine grinding" is not simply decorative, however. It removes a fine layer of metal from the blade faces, making them thinner and more convex, which creates a finely ground but not entirely smooth surface that makes for better food release than a rough or mirror-polished surface. These are the trademarks that make a smooth-cutting knife, and they're born of centuries of observation and learning, and they might be gone if not for the work of the Robert Herder company.

This whole room, the home of the blue-glazing process, represents a phenomenal triumph of memory. Blue-glaze technology almost disappeared with Wilfried Fehrekampf, a *Heimarbeiter* who used to supply blue glaze for the Robert Herder company. When he reached retirement age, there was no system in place—and no apprentices—to capture and salvage what he knew. So Giselheid found apprentices, which wasn't easy, and under the unblinking eyes of two taxidermic birds, they learned to be *Blaupleister*, or "blue glazers." In the days when conditions were worse, there would've been live birds signaling workers to open the windows when they stopped singing because it meant the air had gotten too dusty. Wilfried Fehrekampf has since passed away, but his legacy lives on thanks to Giselheid.

Anatomy of Single Bevel

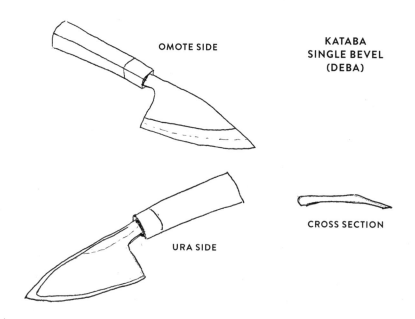

OMOTE SIDE

KATABA
SINGLE BEVEL
(DEBA)

CROSS SECTION

URA SIDE

Anatomy of Double Bevel

RYOBA DOUBLE BEVEL (GYUTO)

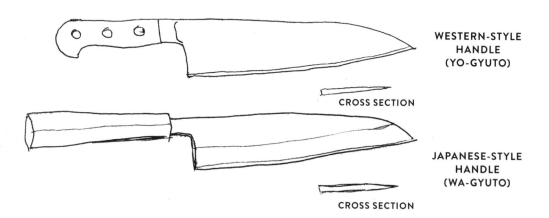

WESTERN-STYLE
HANDLE
(YO-GYUTO)

CROSS SECTION

JAPANESE-STYLE
HANDLE
(WA-GYUTO)

CROSS SECTION

Forging a Gyuto with Shehan Prull

I first learned about Shehan Prull when his mother wandered into Bernal Cutlery. He's a blacksmith in New Mexico who makes high-quality Japanese-style cutlery, and who studied with the master craftspeople at Ashi Hamono, one of the shop's suppliers in Sakai, Japan (see page 102). His style mixes European and Japanese techniques, and he was kind enough to let me invade his workshop and follow the making of a gyuto from start to finish. As he worked, Prull described the process.

We will be forging a *gyuto* from 52100 steel, an American steel that is a low-alloy chromium containing non-stainless steel.

Here, I'm using metallurgical coke and pine charcoal in the forge. Pine charcoal is really the only charcoal you can forge with, [as] it burns hot if given enough oxygen, and produces very little ash, which blows off the top. But it takes work. If you forge exclusively with pine charcoal, you will need an army of minions chopping charcoal for you all day long, so this is why I mix it with metallurgical coke in various ratios.

Once the steel is hot enough, the first step is to shape out the tip onto the end of the bar and then work down the blade from the tip toward the chin. As I go, I'll start checking the shape of the forging against a template.

The forging not only shapes the blade but also controls the heat to form the grain structure of the steel. If you forge too hot or too cold, or heat it up too many times, you will affect the grain adversely. Different steels have different forging temperature ranges; you can work pure iron when it's blistering hot or much cooler, but the higher the alloy content in a steel, the narrower that temperature window, and the more careful you have to be.

I can forge this steel pretty hot at the beginning, but the hotter and longer you heat the steel, the more the grain structure will grow. I want to finish most of the forging in a few hot heats and then gradually lower the temperature as I refine the shape.

After I have shaped out the form, I will cut off the blade where the end of the tang will be.

Now, after heating the blade back up, I will notch in to form the shoulder and draw out the tang. Because the grain in the tang is not as important, it can be forged a little hotter. If I burn it, it would be too brittle and might break, but there's still a wider window of tolerance here.

At this point, the basic shape is complete, and it will take a few more heats to refine that shape, but the focus now is on how the heat affects the grain structure of the steel. Iron has a crystalline structure, almost regardless of how it is alloyed. Both pure iron and iron that's 20 percent alloy will have a grain structure. By controlling the temperature and rate of cooling, I can control the size and distribution of the grain.

Right now, the blade is still a little thick in the spine, and I better take some out now or I will have too much work later on. I am also going to widen the heel, and I want to work while there is still heat in the steel.

The shaping is done now, and I am going to put charcoal—no coke—on the fire and heat it up to do the hammer mark texture, or *tsuchime*, on the blade. I like the hammer marks for the extra food release they provide, but stylistically they also reference the fact that the knife is forged. The black oxide of the *kurouchi*—or "blacksmith's finish"—that's left inside the *tsuchime* provides a little bit of rust protection, too. In traditional Japanese forging, one reason it was left on was for that rust resistance.

I will be heating the steel up to a dull orange to straighten it out and homogenize the grain. Up to this point in the forging process, it was okay to heat just one section of the blade at a time, but at this last stage, I want to evenly heat the entire blade to homogenize the crystals and take out any irregularities or stresses from the spot heats. By keeping the heat low at this stage, I'm keeping the grain structure small.

The pneumatic press I made with square steel tubing takes the heat down to below a glowing heat faster than air cooling does. But this pressure doesn't completely straighten the blade, and there will be several rounds of straightening, or *hizumi-tori*, later. This quick cooling keeps the carbide structure in the grain from getting too coarse or too fine. This steel is capable of going much finer, but it won't perform in the same way if we make it too fine. The grain can actually be seen with the naked eye at this level, although finer points need to be observed under a microscope. It should look matte gray if snapped in half, and a larger grain would be recognizable by the look of chunkier crystals. This steel, 52100, was designed for making bearings, and it can be treated to produce incredibly fine-grained structures that have very high wear resistance. Although a blade needs to have a relatively fine structure to form a good edge, with this steel I find that if I make the grain as fine as possible, it will hold an edge for a long time, but it makes sharpening difficult. And it does not cut as nicely as it does with a little less grain reduction.

Think of grain as soap bubbles: it grows with time and temperature. The hotter it gets or the longer it stays hot, the larger the bubbles get. In the case of steel, each time the metal is heated, new grain boundaries form on the existing ones, essentially multiplying them, so heating and then promptly cooling the steel repeatedly increases the number of grains (per unit volume) of the steel. How quickly we cool the steel affects what phases form and which has the most effect on its mechanical properties. But we'll get into that later.

By forging, by which I mean plastically deforming the steel with a hammer or press, the bands of carbide are brought into alignment with the orientation of the blade. Depending on how the blade is forged and heat treated, this characteristic can be minimized or accentuated, but it is essentially present in any forged, or even hot-rolled, steel. My 52100 blades have a somewhat coarse and visible carbide banding in them. This is not something I set out to accomplish for its own sake, but in pursuing the edge characteristics I wanted, that is how they turned out.

At first, when I followed everyone else's protocol, the knives I made held their edges well but resisted sharpening. They also didn't give me feedback when I cut with them. I don't know how else to describe that quality aside from the Japanese word for it: *kireaji*, or the "cutting feel."

Ideally, I want the steel to sharpen easily and pleasurably. Sharpening is as much a part of owning a knife as cutting with it is, so it's built into the mind-set of making a knife. I could set the benchmark differently, but I'm not looking for how many times the knife can cut a rope or hack a two-by-four. It's a culinary knife.

After forging, the next step will be normalizing in an annealing kiln. The knife is brought up to just below the austenitizing temperature—at which point the crystalline structure changes and reorganizes its elements into the matrix—and then held there for a relatively long time, about three hours. This softens it a bit by breaking down the different crystal structures that will have formed during forging and cooling—some of which are tough, such as pearlite, replacing them with a much softer and more homogeneous structure of small cementite, or carbide, spheres in a ferrite, or iron, matrix.

This type of annealing is done often with a high-chromium alloy steel. A traditional method for annealing is to let the blade rest in ash—especially straw ash—after forging. The ash insulates the blade well, and it cools very slowly. I have an ash bucket, but it's not effective for steels like 52100, so I tend to use it for Japanese-style knives made with carbon steel and pure, soft iron backings.

At this point, the steel is soft enough to be bent with a hammer and cut. This is the phase where I straighten the blade and cut the excess metal from the outline of the knife with the metal shears. Then, I take it to the grinding wheel to shape the blade within about 90 percent of its final form, which some smiths might do after hardening. While it is still soft, the *mei kiri*, or "signature," is done.

Then I'll apply a thin layer of clay to the blade to prevent oxidation and decarburization of the surface while I harden it through another cycle of heating and rapid cooling.

I give my clay slip in a bucket a good mixing and then apply it. I've found a few clays with the right properties here in New Mexico, and I like to use those.

If the clay doesn't come off during quenching, it's not good. And if the clay flakes off in the kiln, it's not good either. The clay needs to come off right at the time of the quenching to be the most effective.

After the light layer of clay has dried, the kiln is brought to temperature to heat the blade. But before I insert the blade, I'll preheat it in the charcoal forge somewhere close to the temperature of the kiln. Having the knives at precisely the right heat before quenching is critical. The temperature must be hot enough to austenitize the steel, but not so hot that we get unnecessary and unwanted grain growth.

Once the blade is up to temperature, I'll dip it in oil to quench it. Most of the clay scours off as it quenches. This rapid cooling from an austenitizing temperature causes some of the carbon atoms that are present in the steel to get trapped inside the cubes of iron atoms as they rapidly transition from a surface-centered organization to body centered. This new body-centered structure is tetragonal in shape and usually called fresh martensite. It is very hard, giving a tool such as a knife the necessary mechanical properties to hold an edge and cut through various materials, but that makes it very brittle and prone to cracking and chipping, too.

After the quench, the blades are tempered for a few hours in the kiln. This process removes the excess hardness that forms during the quench by relaxing the tetragons back to a cubic geometry. To control how much hardness is retained, you adjust the temperature that you bring the steel back up to after the hardening quench. As the temperature increases, so does ductility, but the hardness goes down.

After the tempering, it is time to clean the knife and finish the grinding. First, I clean the clay and oxide from the blade.

To shape the blade into its working geometry, I will start on the large Japanese grinding wheel, the *kaiten mizu toishi*. It's a coarse stone wheel that removes steel quickly, and especially so if the blade is mounted in a *togi* board, which allows for faster work. This wheel was brought over from Japan, and the motor and body were built here.

Next, I will be thinning and refining the shape of the knife, removing the excess metal from the sides of the blade.

For the first shaping with *kaiten mizu toishi*, I'll put in a nice, consistent thickness with a *hamaguri*, or clamshell shape. I'll go from side to side, rather than finishing one side before I move to the next. I'd rather not remove too much metal from one side and then have to correct by removing more on the opposite side. I learned not so much by being told how to do it but by working alongside the Ashi brothers and Mr. Takada. They have two of these wheels there, and I could grind alongside them, although a lot of the time I would grind when they were doing other work and they didn't need the wheels. I learned a lot from just watching how they worked, even though they were so fast.

To check the thickness, I'll flatten the edge a little by running the blade against the side of the stone. Any thick areas will show up as a wider flat spot on the edge.

After I have shaped the sides and established the right thinness, I'll bring the sides of the blade to a series of belts to refine the scratches and smooth out any irregularities. The *kaiten mizu toishi* is very coarse, so the scratches are much deeper than what I want for a finish.

It's important not to overheat or overthin the blade on the belts, so you have to dive right in and keep the blade moving. I start from the tip and work in from there. After the belts, I bring the knife to a flapwheel and then hand sharpen it to finish.

At this point, the blade is finished, and it is ready to be handled. I really like the Japanese idea of the knife and the handle being two separate things. In Japan, handles are traditionally not glued on and would be able to be taken apart for cleaning or replacement.

To set the blade in the handle, I will heat the tang and let it burn its way into the slot cut into the handle. It's very important to make sure the blade is straight before it's driven in. By hitting the bottom of the handle, the tang is driven in and it is set. And there we have it—a finished knife!

Chapter 3

Japan

By some standards, I was a little late to the Japanese knife party. Japanese knives began regularly appearing in Western kitchens in the 1980s, back when most professional chefs were using German knives, which many still do. Those were sturdy and versatile, but when the first professional-quality Japanese knives arrived, they danced circles around the Germans; light, airy, and smooth cutting, they dropped through carrots like a hot edge through ice cream. The *gyuto* started creeping into kitchens, and slowly but surely won converts, but it wasn't until the Internet took off that the *gyuto* did, too.

It was 2006 by the time I slipped my first *gyuto*—the same middling secondhand Misono you met earlier—through that one cabbage and left the church of bigger factory-made Western knives behind (granted, I'll never abandon the well-kept European knives from my flea market days, but they just can't do a carrot like a Japanese knife can).

It's all about the cutting *feel*. Those of us who've come to love Japanese knives like what they can do, but it's the way they feel that hooks most of us—the elegant, easeful way they shake hands with whatever you're cutting. A clean cut and smooth cutting feel are the pinnacle qualities of a Japanese knife. The clean cut actually affects the flavor of your food; a jagged edge cutting through fish flesh will tear cells, releasing enzymes that oxidize and change flavor immediately. The same goes for vegetables. The difference between an oxidized, leaking, musky onion and a smooth, pungent, bright-tasting onion is the difference between a dull, thick knife and a sharp, thin one. In a Japanese kitchen, cutting is cooking, not just something that comes before it.

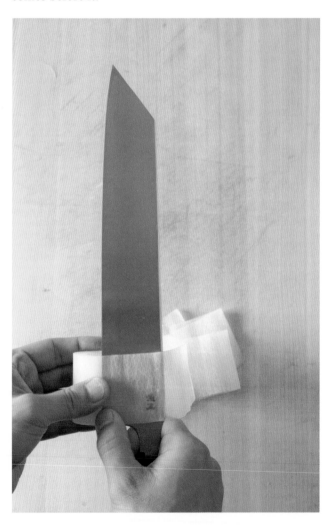

Traditional Japanese Knives

The single bevel is a distinctive feature of Japanese knives, and you'll see it on many of the knives included here. On a single-bevel knife, a wide bevel forms a cutting edge on the *omote*, outside or knuckle side when the knife is held, opposite a concave or hollow ground face on the inside, or *ura*, which has a small, flat bevel called the *uraoshi*. Single bevels accentuate precision and a smooth cutting feel, both focal points of Japanese knife design. Generally speaking, Japanese single-bevel knives are single-purpose knives.

DEBA

The *deba* is a stout, triangular, usually single-beveled blade made for breaking down fish and other seafood. The *deba* dates back to the late 1500s, and was one of the most important Japanese knives until the advent of refrigeration and prepackaged and prebutchered fish made home butchering obsolete. The heel of the *deba* is used for rough work like cutting through the spine, whereas the center of the blade is ideal for refined work like cutting fillets. Although the *deba* is a heavy knife, it is capable of great precision and, when used skillfully, leaves only minimal meat on the bones.

The *deba* comes in a variety of sizes, generally categorized according to thickness and width, that reflects the range of fish eaten in Japan. Shorter, thinner knives, starting at about 100 mm/4 in long, are called *ko-deba*. Thicker, larger knives have a few names and run from 135 to 210 mm/5.5 to 8 in, with the standard range from 150 to 210 mm/6 to 8 in, and the most commonly used between 165 and 180 mm/6.5 to 7 in. If a standard *deba* is made wider, it's called a *hon-deba*. If it is made narrower, it's an *ai-deba*.

MIOROSHI DEBA AND MIOROSHI

A *mioroshi deba* has an even longer, narrower blade than an *ai-deba* and is used for breaking down softer-boned fish like salmon and for portioning and trimming large cuts of fish. When made with a thinner edge and a wider *kiriba* (blade bevel, or inclined surface that starts above the edge bevel), it is called a *mioroshi* and is more suitable for slicing sashimi than for cutting bones. The *mioroshi* sits on the fence somewhere between a *deba* and a *yanagi*.

FUNAYUKI

The term *funayuki* is applied to a wide range of knives that can be single (*kitaba*) or double beveled (*ryoba*) and are often associated with fishermen who, as the story goes, would keep them on the boat for odd jobs or meal prep. They vary in size from roughly 120 to 180 mm/4¾ to 7 in. A double-bevel model has a slim, triangular blade, like a small chef knife, whereas a single-bevel *funayuki* has the same footprint as a small *mioroshi*. One distinguishing characteristic of the *funayuki* besides its small size is the *machi*, or "notch," ground in at the neck on a single-bevel *funayuki*. This differs from the straight tang to neck of a *deba* or a *mioroshi*.

NAKIRI

The *nakiri* is one of the oldest, most continually used knives in Japan. It is a simple, double-bevel, rectangular blade, typically between 165 and 180 mm/6½ and 7 in, though both larger and smaller ones exist. The name essentially translates to "vegetable knife," and it's great for that use and not so great for everything else. The *nakiri* and *deba* were the staples of the Japanese kitchen for hundreds of years and reflect the importance of fish and vegetables as staple foods in Japanese culture. The *nakiri* is either forged with a three-layer blade or a single-steel blade.

USUBA

Designed for vegetables, the *usuba* is a single-bevel, thin-edged knife. Although the steel is often the same type and hardness as that used for a *deba*, the thin edge makes it much more delicate (and better for vegetables). Like several other single-bevel Japanese knives developed during the Edo period, two main styles of the *usuba* exist: the Kanto (developed in the Tokyo area) and the Kansai (from Osaka and Kyoto). The Kanto-style blade is square tipped, with a spine that either ends in a pointed corner or a rounded tip. The Kansai style, also called a *kamagata usuba* because it looks like a sickle (*kama*), has a straight edge and a spine that curves down to meet it.

Both styles of the *usuba* are used for vegetable work, especially for *katsuramuki*, a rotary peeling technique in which a very thin sheet is scrolled off a cylindrical vegetable, such as daikon or cucumber, and then typically cut into an extra-fine julienne. Cutting this way produces virtually no bruising of the cell walls, which means the surface of the cut is smooth and the vegetable won't oxidize like a mashed or bruised slice would.

MUKIMONO

This is a Kanto-style single-bevel vegetable knife that is slightly thinner and usually shorter than the *usuba*. It has the Kanto-style diamond-shaped tip also seen in the *kiritsuke*. This tip style allows for a flat edge and a thin point, similar to the *kamagata usuba* tip but thinner. The *mukimono* is used for vegetable peeling, and its tip works well for detailed work.

YANAGI

Also called *yanagiba* after the willow-leaf shape of the blade, the *yanagi*, which was developed in the Kansai region during the Edo period, is a single-bevel slicer designed for cutting fish for sushi. As with other single-bevel knives, the Yanagi's *ura* side is hollow ground and resists sticking to the slice of fish being cut away. As with other Edo-era knives, distinct Kanto and Kansai styles of the *yanagi* developed. The Kansai style has a slightly curved edge and pointed tip and is a bit heavier than the square-tipped Kanto-style *takobiki*, which is also used for slicing raw fish. The Kansai-style *yanagi* has become the de rigueur knife for cutting sashimi.

TAKOBIKI

This Kanto-style single-bevel slicer, also known as a *takohiki*, is now largely considered a knife for octopus, though it evolved from a square *yanagi*, and many older sushi chefs in Tokyo still use a *takobiki* for slicing sashimi. The name *takobiki* comes from *tako*, or "octopus," and *biki*, "to pull," in reference to the motion used to remove a slice. It is considered a safe choice for breaking and cleaning octopus because it has no point that might pierce the guts as you pull off the outer skin of the tentacles. The *takobiki* is narrow and thinner than the typical *yanagi* and is especially useful for cutting cured whitefish like sea bream or snapper, which are denser and more prone to sticking to a knife than fattier fish that are not cut as thinly. Of the competing creation myths of the *takobiki*, one take is that the Edo-era sushi chefs who sat at their carts selling sushi discovered that a flat blade proved a better tool for cutting raw fish when seated. A curved blade better accommodates the style of today's sushi chefs, who usually stand while they work. Another origin story of the square-tipped *takobiki* is that the local and stationed daimyo and samurai of Edo objected to a pointed weapon-like blade in their presence on the streets of Edo, and the square tip was in deference to them.

FUGUBIKI

The *fugubiki* (or *fuguhiki*) is a thin, narrow version of the Kansai-style *yanagi*, designed especially for cutting the superthin transparent slices of puffer fish (*fugu*). It works well for the sticky, thinly sliced *fugu* in the same way a *takobiki* works for cured whitefish.

KIRITSUKE

A somewhat recently developed knife with a pointed, diamond-shaped tip identical to that of the *mukimono*, the *kiritsuke* is used for the same purposes as the *usuba* and the *yanagi*. It has a slightly thicker edge than the *usuba* and is often used in heavier contact with a cutting board, which wears on a knife's edge. As traditional and ultra-Japanese as the *kiritsuke* appears, it hearkens back to the newer, multipurpose function of the *gyuto* and is one of the few single-bevel knives intended for vegetables and animal protein (fish) alike. The *kiritsuke* is often the tool of the head chef for a *kaiseki* (multicourse dinner) and would not be seen in the hands of a junior cook.

KIRITSUKE YANAGI

As the name suggests, this is a *yanagi* with a *kiritsuke*-shaped tip, originally used in the Kanto area as a knife for cutting sashimi and for doing some vegetable

garnish work. As much as Japanese knives have evolved into specific functions, many sushi chefs will use their *yanagi* to do some of the vegetable prep for their station. Thin cuts of green onion or cucumber are often done with a *yanagi*, and a *kiritsuke* tip allows for detailed work. Although care must be taken not to drive the tip into the cutting board, the knife is also easier to sharpen. The *kiritsuke yanagi* is often a little heavier than the standard *yanagi*, but some chefs prefer this weight. In the case of a *yanagi*, the weight of the knife will help to slice the fish more precisely than the exertion of a lot of pressure will. For this reason, many professionals prefer a longer *yanagi* of 300 mm/12 in or more.

SAKIMARU TAKOBIKI

Another recent twist on an old design, the *sakimaru takobiki* translates to a "round-tip" *takobiki* and represents a design change intended mostly to enhance its theatrical appeal. It has taken the straight blade and square tip of the classic *takobiki* and given it a curved blade and extended the square tip into a sword-like point. The single-bevel *sakimaru takobiki* has the appearance of a small, graceful Japanese sword, and at a sushi bar in front of diners, it has an impressive look. The tip is not particularly useful because it's not sharpened when it leaves the grinders, but the curved edge does allow for a bit more of a sweeping cut, which can be helpful.

UNAGISAKI

Used for eel, the single-bevel *unagisaki* is the most varied of all the Japanese knives when it comes to regional styles largely because of the many varieties of eel caught in Japan and of the diverse local styles for breaking them down. These elongated fish are cut either from the back or the belly, and these different cuts call for differently shaped knives.

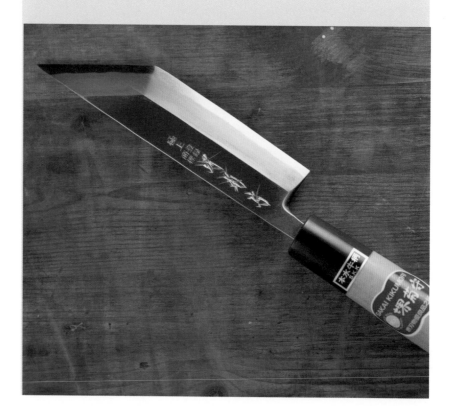

Western-Style Japanese Knives

Western-style Japanese knives are the mainstay of Bernal Cutlery. They have double-bevels, and Western footprints with a Japanese twist. They include two subgroups classified by the type of handle: *wa*, which refers to the Japanese handle, and *yo*, which refers to the Western handle. The knives are otherwise generally the same.

WA-GYUTO AND YO-GYUTO

The *gyuto* is the Japanese take on the Western chef knife, introduced during the Meiji restoration of the 1870s. As such, it can be used very much like a chef knife, though it has a thinner, harder blade that makes it more precise and less tolerant of rough treatment. A heavier *gyuto* is called a *yo-deba*, a thicker version reserved for butchery. In Japan, *gyuto* knives have been mostly relegated to professional kitchens and didn't catch on in home kitchens until after 2000.

The term *gyuto* is often translated as "cow sword," and it was originally designed for cutting beef, but the modern *gyuto* has a broader range that includes vegetables. The first *gyuto* was the *yo-gyuto* (the Western-handled *gyuto*), forged from a single steel with a riveted handle. The modern *yo-gyuto* is distinguished by a nonriveted Japanese-style handle with a three-quarter-length tang, like a single-bevel knife. Although the *wa-gyuto* looks more traditional than the *yo-gyuto*, it came about after the *yo-gyuto*, as did the use of laminated steel blades.

YO-DEBA

The *yo-deba* has the same shape as the *gyuto* but is significantly thicker and designed for butchery. Most commonly used in fish breaking and pork butchery, it is not made for cutting through thick bones. Many old *gyuto* knives incorporate a thick heel similar to that of a *yo-deba* but have a thinner belly and tip.

SANTOKU

The *santoku* ("three virtues"), or *bunka bōchō* ("cultural knife"), is another interesting example of a Japanese knife that seems traditional but actually came about relatively recently. Smaller and more accessible than a *gyuto*, the *santoku* was marketed as an authentic "traditional" knife. It could handle the same vegetables as the *nakiri*, as well as red meat, a new addition to the Japanese diet. The footprint is shorter than that of the *gyuto*, with a steeper curve toward the tip of the spine.

WA-PETTY AND YO-PETTY

The petty knife is a standard accompaniment to the chef knife. It is the Japanese adaptation of the Western utility knife, which evolved from the French office knife, and it became a mainstay in the post–Meiji era knife kit of Japanese chefs. Japanese petty knives are the same size as the Western utility knife, 4 to 6 in/10 to 15 cm, but have a thinner blade. Their uses are similar, too, and they range from very thin and precise to thicker and tougher. Thinner blades are ideal for fine work, while the thicker ones stand up to use as a boning knife. And like the *gyuto*, the first petty knives were mono-steel with Western-style, or *yo*, handles, and the seemingly more traditional Japanese *wa*-petty knives with a mono-steel or laminated blade date from the latter half of the twentieth century.

WA-SUJIHIKI AND YO-SUJIHIKI

The *sujihiki* is the Japanese answer to the Western slicer, or *tranchelard*. Like other Western-influenced Japanese knives, the *sujihiki* was originally made with a single steel and a Western handle, and only later with a Japanese handle attached to either a mono-steel or laminated-steel blade. The *sujihiki* is used similarly to a Western slicer—for cutting proteins with a pull stroke. It is better suited to a number of slicing jobs than the *yanagi*, the traditional Japanese slicer, though it will not cut sashimi as well.

HONESUKI MARU (OR HANKOTSU)

This is the Japanese answer to a boning knife: ground with a right- or left-hand bias and with more convexity on the *omote* side but without the concave *ura* and *uraoshi* of a single-bevel knife. The *honesuki maru* runs around 150 mm/ 6 in and is used in butchery as a heavy breaking knife (especially good for breaking hanging meat, rather than on the table) and for separating primal cuts and working next to bones. It's a tough knife and precise when sharp, but given its stiffness, it has some limitations in trimming and seaming.

HONESUKI KAKU

A triangular poultry boning and breaking knife ground similarly to the *honesuki maru*, the *honesuki kaku* has a convex *omote* and flat, nonconcave *ura*. This is *the* knife for breaking chicken. It offers both precision and toughness at the heel, which is used to chop through joints. The *garasuki* is an oversized *honesuki kaku* made more for breaking than for precise boning.

KO-SABAKI

A bit of shameless self-promotion here: this is a Bernal Cutlery design that incorporates the general size of a *honesuki maru* but with a thinner blade, a docked-off sharp chin, and a larger Japanese-style handle. This knife is designed for the cooks and butchers who complained that their *honesuki maru* wasn't suited for detail work, like trimming and seaming, especially after a few years of sharpening. Most butchers and cooks would move from a Japanese *honesuki maru* to a standard Western boning knife to do trimming and seaming work, but they'd be disappointed with the edge life or cutting feel of the Western knife. So we tried to make the best of both worlds.

The Art and Origin of Japanese Smithing

Knife evolution in Japan was molded by a great many things, but three elements shaped the heart of what Japanese knives were and have become: *tamahagane*, or "jewel steel"; the *tatara*, the furnace that smelts *tamahagane*; and *tennen toishi*, the sharpening stones carved from the mountains around Kyoto. *Tamahagane* is a hard steel with few impurities, and as a hard steel, is unusually easy to sharpen and forge weld to other metals. These properties were born inside the *tatara*, whose bellows, structure, and process harden the steel with carbon and create stretchy, oxygenated bits of slag that make it malleable. And the ability of *tamahagane* to take a fine edge quickly dovetails nicely with the fine grit and microscopically irregular polishing pattern of *tennen toishi*, which could polish a hard steel, including *tamahagane* swords, like no other stone.

Black Sands and the Tatara

Sources of iron ore are sparse in Japan, and the story of smithing here begins with and relies on *satetsu*, the iron sands in the riverbeds and rocks of the Chugoku region. Japan is home to a few different kinds of iron sand, all of which were produced when volcanic stone formed beneath the earth's crust millions of years ago. The two main classes of iron sand, *akome* and *masa*, are found in diorite rock and granite, respectively, the first used for making pig iron and the second for *tamahagane*. In Chugoku, *satetsu* is harvested from riverbeds in the spring and summer, and from the mountains during the colder months to keep the rivers that flow into cultivated valleys pure during crop seasons. To prepare the sands for smelting, the iron from *satetsu*—naturally 2 to 5 percent iron—is collected with magnets and then further sorted by gravity underwater to reach a stronger concentration of iron in the sand (about 60 percent).

To make use of the *satetsu* iron sand ores, the Japanese developed a clay-walled, bellow-powered furnace called the *tatara*. Although its basic components and operation—four clay walls punctured with air inlets where oxygen floods in to feed the fuel and ore—mimic many of the simple furnaces around the world, the product it turns out has a unique purity.

The dimensions of the *tatara* have changed over time, but the basic design has never been altered. During the Kamakura period (1135–1333), the heyday of Japanese sword making, the *tatara* was tall and narrow, but it has since evolved into a larger, open rectangle about 3 ft/91 cm deep and wide and 9 ft/2.7 m long. It boasts two large sets of bellows that feed air through tuyeres on either side, under a 9-ft/2.7 m deep ventilated foundation of ash and charcoal that lets off humidity.

Tataras are ephemeral and must be rebuilt frequently. First, a floor of ash and charcoal is laid. Next, clay bricks are stacked into four conjoined walls and heated until dry. Then comes the smelting: for three days, charcoal and iron sand are pitched into the pit every thirty minutes for about seventy-two hours, until a "bloom" of iron or steel is ready, at which point, the smelting iron has eaten far enough into the clay walls that the *tatara* must be destroyed and rebuilt again.

Interestingly, the *tatara* creates multiple shades of steel with varying levels of carbon within a single cycle. The ore along the sides where the oxygen floods in will smelt differently from the ore in the center, where it is hotter and less aerated. The carbon concentration of the steel—and therefore its hardness—and the level of richly oxygenated iron slag that makes the steel easy to forge weld, depends on where it lies across the furnace floor. *Tamahagane* is usually somewhere in this mix, as is pig iron, and while anything you pull from a *tatara* has its use, *tamahagane* is prized, and in the days of the samurai would have been reserved for swords, while iron and steel plucked from other parts of the pit became culinary knives.

Compared to other irons and steels, *tamahagane* is low in contaminants, especially sulfur and phosphorous, and has a good balance of hardness with tensility, but the oxygen is what gives it its character. When the *tatara* oxygenates the microscopic inclusions of iron slag, they become stretchy and easy to disperse under the strikes of a forge hammer. This is what makes *tamahagane* tough at high hardness, good for forge welding with softer irons, and easy to sharpen quickly compared to equally hard steels. It might be a short list, but taken together, all these qualities bring possibilities and versatility to edged tools that other steels can't match.

Tamahagane takes to forge welding like ink takes to paper and gave knife-smiths an excellent medium for the chisels, plane irons, knives, and edged tools that rely on sharpness as well as durability.

These days, high-quality cutlery steelmakers in Japan have forgone the *tatara* but still use *satetsu* to create good, well-regarded steel that behaves much like *tamahagane*. But without the *tatara* smelters, they can only approximate the exact structure of "jewel steel." Hitachi Metals, Ltd., one of the world's most productive steelmakers, uses *satetsu* for its finer cutlery steel (including the yellow, white, and blue ones you met in the primer on page 28, all of which are made from the same *satetsu* base).

Japanese Knives vs. Western Knives

Where a Western utility knife can take on chicken and leeks, there is a separate Japanese knife for each. Even though each knife is meant to do less, many argue that specialization makes Japanese knives better at what they do.

Because Japanese and Western knives lead different lives, they do not dull or wear down in the same way. Western knives—dodging from chicken bones to slicing onions—require a durable edge made of softer steel that won't crack or chip if the knife is used improperly. Japanese knives are less forgiving, made with a harder steel and the assumption that the cook knows what he or she is doing, at least enough to keep the vegetable knives away from the fish.

A Western knife stays sharp as long as you sharpen it often. A Japanese knife stays sharp as long as you don't misuse it (although of course it'll perform best when sharpest, so don't *not* sharpen your Japanese knife). The first is designed to take a beating and to bend instead of break, and the second is designed to keep its edge and be properly used. And if you ever do destroy a set of handmade Japanese knives by using them badly, as an unnamed Bernal Cutlery crew member might have done in his younger days, you can bet that when your girlfriend takes them back to Japan for sharpening, there'll be a wrist slap and a lecture in order, and possibly a little shaming. In Japan, the customer is not always right.

Japanese knives are harder, but that's not to say they are difficult to sharpen. On the contrary, the core of a Japanese knife is usually a hard steel, but with the correct heat treatment and forging technique, a hard steel can yield easily to a sharpening stone. Even modern knives that aren't true *tamahagane* approximate its properties, which is to say, easeful sharpening and long edge life.

The high-quality steel that's still produced in Japan has been vital to the survival of smithing. Although the *tatara*-made *tamahagane* is a rarity, some of the most popular top-shelf cutlery steel is still made from *satetsu* (iron sand). Hitachi's Shirogami (white steel) and Aogami (blue steel) series are just two examples. Whatever they use, the knife makers that Bernal Cutlery is lucky enough to work with have successfully married traditional techniques with modern technology and tastes. Understanding those techniques is essential to knowing your knife and how to care for it, so we return to the Japanese knife-making center of Sakai.

Sakai and the Evolution of Knives in Japan

Sakai is one of Japan's oldest ports, and it prospered from lively trade with China and Korea during the Muromachi period (about 1336 to 1573), as well as from domestic trade via the nearby Yamoto River. Along with Kyoto and Osaka, Sakai has long been a center of economic power in Japan. By the sixteenth century, it was already known for its *tamahagane* swords, but the spark that inspired the shift to knives came from seven thousand miles to the west.

In the sixteenth century, when the Portuguese docked their ships filled with tobacco, they needed knives to cut those leaves into shreds for smoking. The local swordsmiths adapted and began making long, straight, narrow cleavers to do the job. Japanese farmers started to cultivate tobacco themselves, establishing a domestic need for those cleavers, and business continued to grow. The Tokugawa shogunate (1603–1868) eventually officially recognized the quality of the cleavers, launching Sakai to prominence as the center of Japanese knife making.

If we follow Sakai from this point to the present day, we find new knives emerging at the points where East and West intersect, adapting to the new needs of kitchens as the flavors within them mix and merge. The most popular knife style Bernal Cutlery sells, Japan's adaptation of the French chef knife, was born at a peculiarly strong moment of Western influence in Japanese kitchens. But before we get there, to understand how and why the *gyuto* evolved the way it did, we must go back to late-sixteenth-century Sakai.

The handmade-knife-making industry of Sakai, like those of Thiers and Solingen, has shown remarkable resilience through five hundred years of

wars and fires and political upheaval. In the late sixteenth century, shortly after the tobacco cleavers put Sakai on the map, it was conquered and burned by the daimyo Oda Nobunaga during his bid to seize control and "unify" Japan.

His successor, Toyotomi Hideyoshi, revived Sakai when he built his castle in Osaka, bringing the region into focus as Japan's new center of power and commerce. As part of that campaign, he opened the door to the West and ushered in a wave of Western music, flavors, and fashion that seeped into the region's bones.

It might be a stretch to suggest that a brief window of Western exposure could form the identity of a place, but Osaka still carries the remnants of those years in its diversity, relaxed attitude, and embrace of Western culture. Those open-door years pulled in the seeds of a culture that continued to bloom, even after the Tokugawa regime moved the center of power to Edo—today's Tokyo—in the beginning of the seventeenth century and began a period of isolationism and rapid change, and even after Japan officially closed its borders in 1635. Osaka's cultural tone had been set.

The inward-facing focus of the Edo period changed Japanese cuisine. The Kansai region, an important center of food production surrounded by flat, fertile rice plains, developed a distinctive style of cooking, and Sakai met the order for knives to match it.

During the Edo period, Sakai produced two knife styles—among many others—that established themselves as the mainstays in home kitchens all over Japan: the *deba* and the *nakiri*. Together they covered the main protein and sides of the Japanese diet, fish and vegetables.

The *deba*, which has a thick spine and a triangular blade sharpened on one side, was built for breaking down fish. Its blade is wide and heavy enough to hold up to fish bones, and the single bevel is adept at navigating the steep and shallow dips of a fish carcass. The *nakiri*, which has a small, rectangular footprint and a double bevel with a thin, straight edge, was designed to slice through vegetables without bruising them. For hundreds of years, these two knives have been the heavy hitters in the Japanese kitchen.

By the 1720s, Sakai, with about one million people, had grown into one of the world's largest cities, and the shogunate had installed a rigid new feudal system and centralized power that took daimyo (landholding lords) off their land and moved them close to the capital. The move quelled any chance of an uprising by keeping the resistance within the shogun's view.

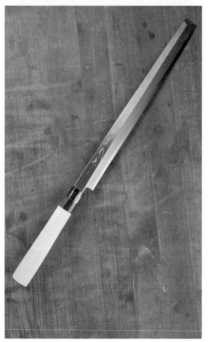

The rules were strict, but they made the Edo period a relatively stable and peaceful one. Those conditions saw the infrastructure blossom and the literacy rate rise to one of the highest in the world. But when the daimyo left home and flooded Edo, the ratio of males to females tipped the balance both in the city and back home. Now Edo was filled with wandering, hungry men—dislocated daimyo living far from home and far from home-cooked meals.

In time, the need to feed this lopsided, famished population inspired the first Japanese restaurants. As cooking migrated outside the home, food changed. It was no longer limited by ingredients, space, or other constraints of the home kitchen. Street vendors and restaurants could expand beyond what had been practical or possible before, giving way to much of the food that modern Westerners recognize as quintessentially Japanese, such as sushi.

The first sushi, which dates back to the Yayoi period (300 B.C.E.–300 C.E.), was salt cured and stored in rice for many months, which turned the rice to paste and fermented the fish. In the 1500s, fermentation stopped short of turning the rice pasty, and the fish was eaten with the rice. During the Edo period, a sudden abundance of cheap rice vinegar made from sake lees did away with fermentation altogether. Now the fast-moving life in Edo enjoyed a new dish: cooked rice soaked in vinegar and topped with *nigiri*-style fresh fish. The rice would have been pink from vinegar and, according to some scholars, the pieces of fish were two or three times the size of what is now common. Refrigeration had yet to be invented, so the fish was still treated to some vinegar, salt, and soy sauce.

On the heels of Edo's new cuisine came the new knives to prepare it, which, like the sushi, swaggered their way into the Japanese culinary canon. Enter the long, thin *yanagi* for fish and the stockier but thin-edged *usuba*, which was similar to the *nakiri* in that it was meant for vegetables but capable of much more intricate work. And just as regional food styles developed, so did a distinctive difference in the knives those regions used. Knives in the Edo-centered Kanto region were made with a square tip, which we can see most clearly on the *yanagi*, while knives in the Kansai region, home to Kyoto and Osaka, favored a pointed tip. These differences were often more stylistic than functional.

Open Borders, Changing Tastes, More Knives

In 1854, Commodore Matthew Perry made a mighty showing of naval strength just off the shore of Japan and suggested that Japan open its borders by the time he returned the following year. The Tokugawa regime acquiesced, sparking a hornet's nest of brooding internal conflict that weakened the regime to the point that when young Prince Mutsuhito took power in 1868 following a coup d'état, there wasn't much to fight against except the ossified remains of the shogunate.

Emperor Mutsuhito moved the imperial palace from Kyoto to Edo, which had been renamed Tokyo, reclaiming power from the decentralized shoguns, reestablishing the preeminence of the emperor, and closing out the Edo period. During his forty-four-year reign, known as the Meiji era, Mutsuhito brought radical reform to Japan and restructured the country into a modern nation-state. Emperor Meiji (*Meiji* means "enlightened rule") surrounded himself with the young samurai from regions that had been part of the coup. After a few bloody uprisings failed, a modern Japan settled into place with no room for old class systems or for samurai and their topknots and swords. Once again, the doors to the West were open, and the Meiji society embraced what came through—especially the food.

On January 24, 1873, Emperor Meiji ate beef. Today that sounds like the windup to an old-fashioned dirty joke, but in the newly minted Meiji era, it was a serious move. Up to this point, red meat consumption had been banned for hundreds of years by edicts in both the indigenous Shinto faith and Buddhism. Fish were widely eaten, but for anyone alive under the Emperor Meiji, this was the first time in their lives that something with four legs had been butchered and cooked.

Emperor Meiji, with Commodore Perry's threat not far from memory, campaigned for a new and more fortifying diet, believing that eating meat would strengthen Japan, empowering her to catch up with the powers that encircled her and to grow—physically and otherwise—beyond the long shadows of the Western gunboats that lurked nearby.

But regardless of what brought it, Western food had arrived. The chefs behind it moved in and set up shop. The Tsukiji district in Tokyo became a hotbed of Western food, with the Tsukiji Hotel, the first Western-style hotel in Japan, opening its doors in 1868. It boasted a restaurant helmed by Louis Beguex, a French chef, and more hotels like it soon opened throughout the nearby Ginza district. Their presence sparked a movement toward Western flavors and styles and the beginning of *yōshoku*, Western-influenced Japanese cuisine.

The spotlight on Western food caught Western cooking equipment in its glow. Until now, the traditional roster of *deba*, *yanagi*, *usuba*, and *nakiri* covered all the demands of the Japanese diet. There were no tools for red meat. Japanese chefs looked into the kitchens of Tokyo's French hotels and Western restaurants and found the perennial all-stars of the Western chef's tool kit: the chef knife and the office knife. And just as the cuisine of the West had been adapted for Japanese ingredients, so, too, was the chef knife transformed to suit the Japanese kitchen.

Enter the Gyuto

The profile of the *gyuto* blurs the lines between East and West, like a blueprint of the cultural mash in Japan at the moment it was created. So, too, does its name, which is derived from *gyu*, or "cow," and *to*, or "sword"—or more accurately, "cow blade." The long blade and broad heel resembled and functioned like a chef knife. At this point, the shadow of Commodore Perry's black ships belching dark smoke into the sky above Tokyo Bay in 1853, with a Hobson's choice hanging from the tips of their guns, still hovered over Japan. The size and threat of foreign power loomed at the horizon, and the Japanese were compelled literally to "beef up" for the betterment of the country. Westerners were larger, their navies mightier, so the Japanese sized up their diets to match. The move to modernize and Westernize was, at its root, about survival: catch up or be swallowed.

The *gyuto* was cleverly marketed (and named) to evoke a bold panache. It was a muscular knife, a beef-eater's blade. They knew as we know now that a name goes a long way toward selling something (just ask the Chinese gooseberry or the Patagonian toothfish, or their alter egos, the kiwifruit and Chilean sea bass).

My own suspicion is that the *gyuto* was named and marketed to evoke brawn and to answer that national yearning for muscularity and strength. I see a Napoleonic branding message: No longer will you be small and inferior. Now you will be the wielder of the sword—the sword of the beefy! Buy one today! In the end, honestly, the name given the *gyuto* may only be a practical reference to the knife's function.

The Santoku

The *gyuto* became a staple in Japan's professional kitchens, but it couldn't gain traction with home cooks. In the early twentieth century, the *nakiri* and *deba* were still holding strong in Japanese kitchens because they were small, familiar, and easy to use. Still, the times and technology were changing. Refrigeration arrived, bringing massive upheavals in the way food was bought and prepared. Where households used to buy their fish whole and butcher them at home with a thick-spined *deba*, now they had the option to buy already-portioned cuts that kept well in the fridge. The redundant *deba*, its traditional job now outsourced, sat in a drawer collecting dust.

Some cooks replaced the *deba* with a small *yanagi*—slender, pointed, and designed for sashimi. But home cooks were still at a loss when it came to meat, and the *gyuto* of that time was too large and unwieldy for preparing small portions at home. They needed something smaller, something new, but not so new that it couldn't be trusted.

Cue the *santoku* (also called the *bunka bōchō*): shorter and more approachable than a chef knife, with some enticingly familiar lines and curves. Its length and blade thickness matched those of the *nakiri*, but it had a slightly curved edge that curled up to meet the downward sloping spine. Where the *nakiri* was a bit clueless with meat, the pointed *santoku* could find its way. It was not a knife for breaking down half a cow, but no one was doing that at home anyway. Meat was growing popular, though not in a let's-tuck-into-a-bloody-steak kind of way. Portions were modest, and meat was usually cut before cooking.

The names *santoku* and *bunka bōchō* were carefully calculated to convey two dueling concepts: Western modernity and a respectable connection to Japanese culture. *Santoku*, or "three virtues," evokes some kind of old wisdom or ancient moral paradigm, just as *bunka bōchō*, or "cultural knife," implies a down-to-earth, authentic, farmhouse practicality. Either is a brilliant marketing ploy, appetizing bait for any modern home cook who wants to be up-to-date but not so neglectful of tradition that *obāsan* (Grandma) would feel uncomfortable spotting one in the kitchen. In the eyes of an *obāsan*, the virile muscle of the *gyuto* might seem like a meaty, offensive homage to the West that left respect for tradition behind. The *santoku*, on the other hand, looked a lot like the past.

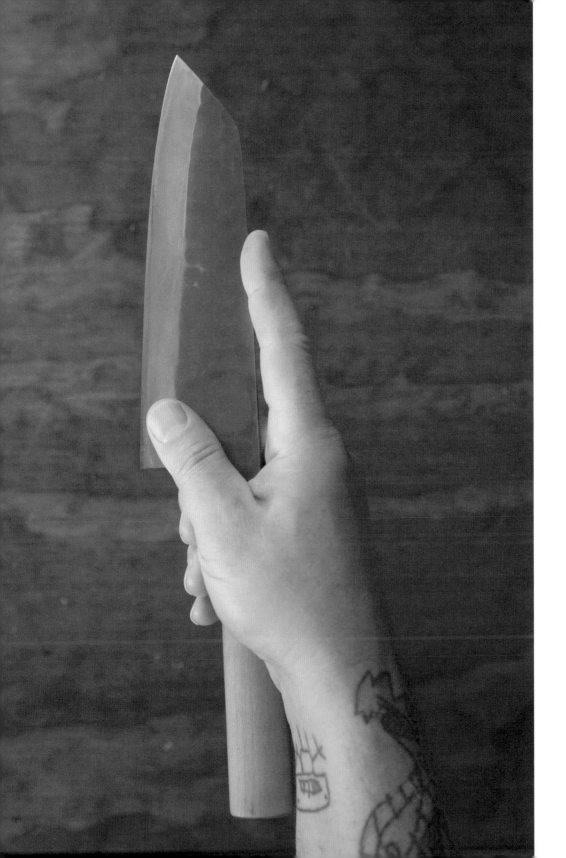

The People

When I first visited Japan in 2014, I went to meet the craftspeople behind the knives we sold. I wanted to know everything about them, and what kind of tricks at the forge or the grinding wheel made a knife cut a carrot like it was butter. I visited again in 2016 to dig a little deeper and to spend a little more time in their workshops.

Most of the techniques I found were hundreds of years old, but the craftsmen who practiced them were fluid, ingenious, and adaptive. Many of these knife makers—my favorites, really—are masters of the tension between the old and the new, using traditional techniques as a framework for creative experimentation. Each craftsman had his own way of relating to tradition, but the thrust of everyone's workshop philosophy always seemed to be the same: get to work and be inquisitive.

I chased the ancestral chef knife in Europe, and in Japan, I had plenty of questions about its Japanese successor and the best-selling knife in our shop, the *gyuto*. Where was it first made, and what did it look like? The oldest *gyuto* I had ever found dated back to Tokyo in the 1950s or 1960s, and I had strong suspicions that the first one appeared long before that. There are short versions of the story, which are neither satisfying nor plentiful, and as I did my own sleuthing, I realized that most of these tales lead back to the same place, and so back I went—back to Sakai.

ASHI HAMONO

On the north end of Sakai, near the Yamato River, down a lightly industrial corridor, is the Ashi Hamono workshop. The shop is about three times the size of an American two-car garage—too small to call a factory—and without a wedge of wasted space. *Kaiten mizu toishi* (grinding wheels) measuring about 4 ft/1.2 m sit cheek by jowl with other smaller grinders, while hulking mint-green machines stand heavy against the walls. In the front hangs a thick-gauge wire sculpture of a *gyuto*. Mr. Ashi is in his seventies, tall and soft spoken, and his eyes are quick and kind. Mr. Takada, his foreman though younger, moves and works with the liquid ease of someone who's been doing this for many years. They are a relaxed and good-humored pair, running one of the most unusual workshops in Sakai.

In Japan, becoming a traditional craftsman is not a road for the cash obsessed. It takes hard work and many years to become proficient, and the earnings are modest—even more so as the years have gone on. Mr. Ashi's father was a traditional knife maker who forged traditional knives—*deba*, *yanagi*, and *usuba*. But the company has made a name for itself producing the *gyuto*, elevating the knife's design and quality with a sensibility that mirrors the artful but understated facade of the factory itself.

Stamped knives—literally stamped by machine from a sheet of metal—don't carry the stature that forged knives do, and companies that deal in forged knives will be quick to tell you as much. Typically, those companies are right. Forging knives breaks up the carbides in the steel and refines the grain, and stamping gets a bad reputation because it's usually part of an industrial factory's attempt to cut corners. With careful heat treatment and special attention at the grinding wheel, the factory restores the touch of the hand and the qualities of a good knife. Ashi knives might not hold an edge as long as many others, but they undergo a heat treatment that makes them amazingly easy to sharpen with Japanese stone. And perhaps most important, you'll be hard pressed to find knives that cut better.

The workshop's bread and butter is the knives they make for other brands, stamped and sold with someone else's seal (a common practice known as original equipment manufacturing, OEM). But the knives we sell at Bernal Cutlery are Ashi Hamono's own brand, and I was very excited to get my foot in the door just before the workshop stopped taking new wholesale customers. They were some of the first knives the shop imported directly from Japan, and they've quickly gained a following in the San Francisco Bay Area and beyond.

In Sakai, similar to Solingen and Thiers, it's unusual to find knives made from start to finish by a single craftsperson. But Ashi Hamono keeps about 90 percent of the process in its own factory.

Most of the blades are stamped from either Swedish stainless-steel or Japanese carbon-steel sheets and then roughly ground by hand. Next, the steel is heated, cooled, and annealed, and then it heads to the *kaiten mizu toishi*, where the knives are secured in a wooden plank (*togi* board) and pressed against a quick-spinning, coarse wheel. As sparks fly and the wheel whips water against the wall, a slightly convex, tapered blade emerges under the practiced hand of Mr. Ashi or Mr. Takada. After this, the shape is refined on a series of belts and flapwheels, and handles are installed by the small handling team upstairs.

To the uninitiated, these knives look like run-of-the-mill Western-style Japanese knives. But in the hand, they are amazingly light and smooth cutting enough to win a loyal following of professional cooks. In our shop, they are the benchmark for a "smooth cutting feel."

Ashi Hamono is a fascinating blend of the old and new. It takes industrial techniques and refines them with a craftsman's touch into the light, convex, smooth-cutting knives that one would never expect from a stamped knife. But I'd also come to the workshop with a mission: to pin down the first *gyuto*. If anyone knew who made the first *gyuto*, I thought it would be Mr. Ashi. He told me the whole story over a few cups of tea.

The History of Ashi Hamono and the First Gyuto

ASHI: Before the Meiji era, there may have been some [chef knives] brought in from abroad, but I don't think there were any made in Japan. Tokyo was the first place in Japan that made the *gyuto*, and the man who made it was a *katana* [sword] blacksmith from Osaka. It started in Meiji 5 [1872], when Minamoto Masahisa—another *katana* blacksmith—became a *deba* knife blacksmith, while his second son, Nakamura Tetsujirou, opened a store near the Tokyo fish market in Nihonbashi, as well as a factory in Kanda-Kajichō. He named himself Azumaminamoto no Masahisa, and started a Japanese knife-production business.

In the middle of the Meiji era, he sent a blacksmith named Sumida Taro to the West to learn Western blade-manufacturing techniques. When he returned, he brought back information about new Western knife-making technology and used it to begin producing the *gyuto* along with other Japanese knives. Another man, named Masakane-san—also known as Hori Kenzaburo-san—also learned under Azumaminamoto no Masahisa. Masakane-san went on to make some of the more influential *gyuto* knives of the twentieth century.

In the beginning, there was a factory in Hodogaya [near Yokohama], and later a larger factory in a place called Gondazaka. I was able to do a tour of that factory, so I know how they did their work there.

In terms of the first *gyuto* made in Sakai, it was actually not in Sakai, but in the next city over, Matsahura, where the people thought the water was good for knife making. The year was Meiji 35 [1902], and the first *gyuto* in the region was made there by Sano, also known as Minamoto no Masamori, who had apprenticed under Azumaminamoto no Masahisa.

When I started making knives, when people thought of the *gyuto*, they automatically thought of Masakane—that's how famous the brand was in the 1960s.

In 1932, my father moved from Toyama Prefecture to Sakai to apprentice to a knife maker. Twelve years later, he started his own business and altered premade knives. Then, in 1956, he began forging his own knives from surplus metal used to make shovels. One day, he noticed his competitor, another knife maker, driving down the street in a brand-new car. My father knew that this competitor had just begun making the *gyuto*, and that the car and the knives probably had something to do with each other. Up to this point, my father had been making only traditional knives: *yanagi*, *deba*, and *usuba*. As the gleaming car whizzed by, he decided that Ashi Hamono would make the *gyuto*, too.

In 1966, I joined the business, and in 1971 I began making Western knives myself. In the beginning, we didn't understand how to select steel or manipulate it. We had mishaps, but gradually we started to get the hang of it. We bought a metallurgical microscope and a pentrometer that measured hardness and then continued to research the process. We are still growing and learning, but that's how we've arrived where we are.

KYOTO

From Sakai, I headed north. The mountains around Kyoto are home to *tennen toishi*, the natural finishing stones that helped to shape the history of Japanese smithing.

Tennen toishi are composed of the compacted skeletal remains of single-celled radiolarians that lived their short lives in the sunny surface waters of the Paleozoic seas hundreds of millions of years ago, settling to the calm seafloor to form a silty mud. This undisturbed mud lithified, and then 175 million years ago, as the Pangaean supercontinent broke up, it began its travels to modern-day Japan as the floor of what is now the Pacific Ocean. As the volcanic island chain of Japan formed, scraps of sediment from the ocean floor were scraped up onto the islands, along with the remains of our Pangaean radiolarians.

Rocks made from these skeletons are not exactly rare (I am sitting about 20 ft/6 m above the chert of the Franciscan formation, made of millions of radiolarians, as I write this in San Francisco). What is unusual about the *tennen toishi* found only in a narrow belt of the Tamba formation around Kyoto is that they did not fully crystallize together into a solid, flinty chert, or turn to a crumbly clay. The best usable seams are rare, and to be effective as sharpening and polishing stones, these rocks must not have excessive compaction, mineralized veins, or too many fissures.

But finding these stones at the source was not easy. The most famous mines have been closed for decades, along with most of the shops that sold the stones. The source of stones is disappearing, but it's the decline of the role of carpenters in the building trades that has changed the *tennen toishi* business in Japan. Carpenters used to be the center of the building trade, but changing building codes, technology, seismic safety regulations, and architectural taste have all but driven out the need for chisels and plane blades, and with it, demand for natural stones.

Thankfully, there is still a dedicated group of natural-stone users and advocates in Japan (and overseas) keeping the demand and enthusiasm for natural stones alive. I'd come to Kyoto to see one guardian in particular, Yozo Dobashi. With his two sons, he operates the last full-time mine in the Tamba region, home to the famous *tennen toishi*.

YOZO DOBASHI

I arrived early by train to Kameoka, where the main-street shops rub elbows with small, meticulously tended fields of the famous Kyoto *negi*. The *negi*, or green onion, is not treated casually here, not like a garnish. *Negi* is a spotlighted protagonist, because it's absolutely amazing. Its tender stalks are sweet and vegetal, worlds apart from the fibrous, sharp-flavored things I've always known as green onions. But I was here for the stones.

Mr. Dobashi, all smiles and warmth, met me at the train station. He's a *tennen toishi* evangelist, a one-man marketing team and sales rep. We drove to his home and workshop down the mountain from the mine, winding through a rural area with little clusters of old homes that used to be full of natural-stone miners, he said. Now they're all gone. His mine almost disappeared with them, but then, in a funny twist, the internet was born and opened his access to a huge customer base in Japan and overseas. The ancient Pangaean rocks of the Tamba terrain started trending, if only a little.

In his home, which doubles as a workshop and stone museum, we rolled up our sleeves and tried a few blades on a dozen or so stones, then we piled into a little pickup and headed to the mine. At a bend in the road, a centuries-old sugi cypress sat next to a shrine, and tall, narrow pines clung to the steeply sloping mountains. At the top of a muddy trail, we came to the mine, a small hole in the side of a mountain. A generator coughed to life in a corrugated-metal shack, and an extension cord peppered with lightbulbs trailed away from the rickety structure into the mine proper. It's one of the last few active *tennen toishi* mines, but there wasn't much life to be seen.

The mine is a wet cavern with striated walls of ocher, gray, and white—the layers of compacted radiolarians that died and dropped to the tropical seafloor millions of years ago, accumulating at a sloth-like pace each century. The awe set in. It took tens of thousands of years to grow the stones that we scrape away at every day in the shop. That wet slurry that broke off as we scrubbed knives across their surface? That was millions of protozoan skeletons unmooring from the stone where they've long been entombed, released now and picking away at the steel.

After a scramble back down the steep mountainside and a bumpy ride back to the workshop in the tiny *keitora* (light truck) loaded with freshly mined stones, we hunkered down with a dozen or so to test them out.

Someone asked me once what I would grab if the store was on fire, and at first I couldn't answer. Later I realized I would take a stone that I got that day from Dobashi-san, an old-stock Tamba mine stone. That, and my hat.

SANJŌ

My next stop was Sanjō in Niigata Prefecture, on Japan's west coast about midway up the northern curve of Honshu. The history of blacksmithing here began by necessity during the Edo period, when the city's main rivers, the Shinano and Ikarashi, periodically flooded the place and devastated its infrastructure. The government commissioned the local farmers to pitch in by making nails to support the rebuild, sprouting a diverse metalworking tradition that grew to include teapots, scissors, woodworking tools, and, of course, knives. In a world that's grown less appreciative of all those things, Sanjō is holding strong.

Compared to other knife-making centers, Sanjō is uniquely remote, but oddly this has worked in its favor. For a long time, hardware merchants who wanted to sell their wares outside of Niigata had to travel a great distance by foot and carry their goods on their backs. It was a long trip, but as traveling voyeurs, they picked up what tool and knife makers in other towns were making, and what people were buying. They cherry-picked the good techniques, styles, and ideas and brought them back to the blacksmiths in Sanjō, who wove it all into their production. Eventually, Sanjō made its name in carpentry tools, from hammers and pliers to saws and planes. But the years since the early 1990s have not been kind to carpenters, who have struggled to stay relevant amid the changing building regulations in Japan and to outshine the massive imports of cheap tools from China. Some tools took a beating, but the knives of Sanjō are still in high demand.

Sanjō knives fall everywhere between industrial and painstakingly hand-made, and the labor is more centralized than in Sakai. The knife makers here do their own smithing, grinding, and handle fitting (though handle making is still left to specialists) in-house, and many of the smiths belong to an informal trade association. The apprenticeship system is alive and well, and the good smiths are swamped with orders. Some say the biggest challenge for a Sanjō knife maker is knowing when to say no.

The industry survives this way because the city has taken care to nourish it. As in Sakai, traditional artisans are aging out in Sanjō, but the city took official action here to address the impending loss, promoting traditional crafts in schools and incentivizing apprenticeships in blacksmithing. Bernal Cutlery is happy to have relationships with a number of the older craftsman in Sanjō as well as with the younger generation, some of whom are just hitting their stride.

I came to Sanjō to visit four of our partners: The Wakui family, in their workshop on the bottom floor of their modest house, and a little farther out of town, in a corrugated-steel industrial corridor, the Yoshikane and Hinoura workshops. At the very edge of town, where the plain ends and the mountains begin, there is the home and small workshop of Shigeyoshi Iwasaki.

SHIGEYOSHI IWASAKI

Shigeyoshi Iwasaki is a retired Sanjō smith famous for his Japanese-style razors, but he's best known locally as a *tamahagane* scholar who brought modern science to traditional Japanese smithing. He introduced metallurgical microscopes to traditional techniques, giving the smith access to a complex understanding of steel that wasn't available before.

Many of the well-known smiths in Sanjō found their start apprenticing with him, but his most significant impact might be strengthening the standard of transparency and education in the Japanese smithing world through his willingness to share his secrets and findings. The research he contributed is dwarfed only by the generosity with which he shared it, and the cultural movement toward sharing knowledge openly is Iwasaki-san's gift to the smithing world in Sanjō and beyond.

It started with his father, Kosuke Iwasaki, a college professor who specialized in ferrous metallurgy and studied the old steel used in Japanese swords. At the end of World War II, after a stint studying heat-treatment engineering at Tokyo University, Kosuke Iwasaki and his family moved back to Sanjō, his hometown. The wake of the war left Sanjō without a peacetime industry, but the recent university student had an idea: What about cutlery? What about making high-quality knives better than what was available? If Sanjō made excellent knives and took them to the export market, the transition into peacetime could be smooth and prosperous.

And so, he started making *kamisori*, or straight razors. They were not the novel kind that wannabe Bay Area beatniks keep next to their dusty antique typewriters, but the real kind, from the time when shaving with a straight razor was a worldwide norm. Back then, it was a promising move: the combination of high-quality, traditional Japanese forging and a global desire for a clean shave.

Kosuke Iwasaki sent his son, Shigeyoshi, to study under one of the top three smiths at the time, while he brought his own metallurgical research to the smiths of Sanjō in hopes that a deeper scientific understanding of forging and its materials would help them improve their own practice and elevate the whole of Sanjō knife making. When Shigeyoshi finished his training, he returned and began making razors.

Despite his education, his first razors got some complaints. Feeling insecure and doubtful, Shigeyoshi made his razors harder, with a keener edge, and partnered with his father to devise a new way of using a hard, carbon-rich *tamahagane*. He took that scientific, analytical eye to the natural sharpening stones of Kyoto, studied the edges they left, and refined the industry's understanding of how the stones interact with steel, especially *tamahagane*. If he had simply known that the barbers who used his razors had bad stones and couldn't sharpen them well, we might not have the legacy that Shigeyoshi has left.

The production of razors in the Iwasaki workshop has since passed on to Mizuochi Ryouichi, Shigeyoshi Iwasaki's former apprentice, who sees to all the phases of smithing, grinding, and sharpening. But little has changed: the razors are still produced using the techniques developed by Shigeyoshi Iwasaki, who remains a regular presence in the workshop.

Both of them were in the workshop when I visited. Ryouichi-san is a patient teacher and walked me through a lesson on sharpening a *kamisori* razor on synthetic and natural stones. Both men, in classic Iwasaki form, are endlessly generous with their time and knowledge. And loyal to the family legacy, Shigeyoshi Iwasaki continues to carry his father's torch of transparency in Sanjō. He was instrumental in forming the Sanjō Kaji Shudan group of smiths who volunteer in far-flung areas of Asia and Africa, bringing the town's trade secrets out of the shadows to places beyond Japan.

It's hard to capture the importance of Shigeyoshi Iwasaki's dedication to sharing his knowledge, and the amount of respect the cutlery world holds

for him. I was told as much by each smith I visited. His joy and generosity is truly infectious, and ironically, in his effort to raise the tide and lift all ships together, he's made a bigger name for himself than he ever could have otherwise.

TSUKASA AND MUTSUMI HINOURA

The Hinoura family has been in the Sanjō cutlery business since 1915, making sickles and then *deba* knives, and after the war, Japanese hatchets. Three generations later, the workshop is in the hands of Tsukasa Hinoura and his son, Mutsumi, who together make knives that are both rustic and refined in appearance and in use.

Tsukasa started forging when he was twenty-two. He apprenticed with Shigeyoshi Iwasaki and, as any true Iwasaki pupil would, has amassed piles of metallurgical microscope photos and a phenomenal amount of research on the structure of steel forged at different temperatures and heat treatments. He has even collected samples from knives from all over Japan.

People used to tell Tsukasa that he was wasting all that time at the microscope, but he is his own biggest critic and wanted to master his craft. He knew that understanding the metallurgy through a microscope was key to honing his understanding of what the metal was doing, and now he can guess at what the naked eye can't see when the hammer strikes a glowing billet.

Before he started forging culinary knives, Tsukasa was well-known for his outdoor knives—blades for clearing brush and hunting. Nowadays, he has a loyal following for the superbly forged, carefully heated culinary knives, many of them marked by his unique swirling style of *jigane* (cladding).

His *warikomi*-style forged kitchen knives have cladding wrapped around the *hagane* (hard inner core) like a taco, rather than pressed on both sides like a sandwich. Some of his knives bear the mesmerizing "river jump" pattern of *gokunantetsu* (pure soft iron) and mild steel forged in a gently undulating twist. It snakes onto the back of the spine, dripping down across the other face of the blade. Beauty is important to Tsukasa, though not for its own sake. Although the pattern has a subtle, transfixing flow, the real artistry is in the functional harmony of the steels.

His technique hinges on the wisdom of his mentors, who taught him to prolong the forging and control his heat cycles carefully. He trained within the classic blacksmithing tradition of Sanjō, and while he's faithful to tradition, he's not dogmatic about it. He splashes water on his anvil to remove the scale that forms on the steel during forging to protect the knives, a messy practice that most smiths forgo.

Tempered hard, Tsukasa's Shirogami (white) carbon steel has a crisp feel. It's remarkably tough given its hardness (remember that hard can mean brittle), and it can take an incredibly sharp edge. His knives are ground with wide *kiriba* bevels that are easily thinned during sharpening. In Japan, where sharpening is an implicit, practical piece of daily knife ownership (unlike in the West, where sharpening is avoided like a chore), steels are often selected based on how well they shape up on a sharpening stone. Tsukasa is partial to carbon steels like Shirogami, a hard, two-ingredient steel that takes to a stone nicely. Shirogami is a kind of litmus test for any smith, much harder to forge well than a knife with highly engineered, easy-to-work-with alloys.

Tsukasa has been training Mutsumi to smith since his college graduation in 2001. Young and well regarded in the smithing world, Mutsumi is now coming into his own. His knives are much like his father's, of course, but different, too. Each of them grinds a dramatic, distinctive taper from the beginning of the spine to the tip, but Mutsumi's knives curve a little more, and Tsukasa's are flatter, if only by a hair.

Temperature is at the core of their techniques, and their ability to control it well enough instinctively to create a fine grain and a crisp but workable steel speaks volumes about the experience, knowledge, and skill of both father and son. But building that instinct takes time. Mutsumi tells me that some sixth senses are hard to build, and in the beginning, it wasn't easy to look at a flame and know its temperature.

And instinct can't be taught, either. It's a matter of doing—over and over and over again. Mutsumi makes it look easy, but when I ask him to name the most difficult thing to learn, he immediately says, without a hint of irony, "everything."

YOSHIKANE

Yoshikane knives were some of the first I ever ordered directly from Japan. They're a confounding mix: substantial but smooth cutting. And the smoothness of their cutting is only eclipsed by the smoothness of their sharpening. The steel is hard but soft on the stone, and the knives take an amazingly fine edge with a quarter of the work you'd expect.

The workshop was opened in Sanjō in 1919, by the same family that runs it today. As someone who spends days at a time at a sharpening stone, I wanted to know everything about what makes Yoshikane knives so easy to sharpen. According to Kazuomi Yamamoto, a fourth-generation knife maker, there is no single reason. A combination of the tempering and the geometry, the quality of the materials, and the skill of the hands is what makes a Yoshikane knife.

But that felt like a vague answer to me, so I pressed him for more. Good-quality steel is important, he said, and they spent years experimenting to find out what they liked and what they didn't like. There is the thorough forging, and the careful tweaking of the temperature here and there. And then many of the blades are tempered not once, but twice, after annealing. This helps to tame a hard steel that would otherwise resist a sharpening stone.

And then there is the geometry: keeping a thin, low-shouldered bevel that rolls nicely on a brick of *tennen toishi*. Like a typical Sanjō workshop, most of the work happens in-house here.

Yamamoto-san captures the Yoshikane philosophy in those two phrases you met in the very beginning of this book: *sessa takuma* and *shoshin wasuru bekarazu*. In other words, evolve your skills through friendly competition, think like a beginner, and remember that learning is never finished. There is no room for egoism in knife making, because without community and collective learning, knives would be made in the dark.

WAKUI

The Wakui family workshop is the ground floor of a wood-frame house on a street corner in a residential neighborhood in Sanjō. If Toshihiro or his father is not forging, you'd never know what happens behind the facade of this modest building, with its slatted wooden windows, big pots of geraniums, and fish tank by the front door.

Wakui-san makes excellent knives with a crisp yet buttery cutting feel, and his forging and grinding are always on point. The heat treatment and the smoothness of the blades on the sharpening stone speak to the time he spent apprenticing with Kazuomi Yamamoto at Yoshikane. In 2004, when things were looking bleak for his family's hardware-forging business, Wakui-san started working triple shifts to build his foundation: in the morning, coarse grinding (*aratogi*) at a large knife maker; in the afternoon, a few hours at the family business; and at night, an apprenticeship with Yoshikane.

Bit by bit, he collected machines and wheels and tools, and after four years, he opened his own operation. But at first, the geometries of his knives were off; he needed practice, more time to experiment, and some mentorship here and there. Luckily, he was a smith in Sanjō, where the culture gladly lifted young smiths through the ranks. Young smiths like Toshihiro Wakui—and there are a number of them—are dedicated to soaking up what the older generation has to give.

Chapter 4

Sharpening

Sharpening techniques share a certain common strategy, but there are many ways to the finish line. The technique I will be describing here has become my style, but I cannot truly claim ownership of it. It has evolved from my own experiments at the kitchen sink, from a decade of passing all kinds of knives over the stones in the shop, and from sharpening with employees who bring their own experience and perspective to bear on our method. With hours at the sharpening stone, the Bernal Cutlery workers invariably develop their own techniques and tricks that they, in time, teach me. As with any functional art, the proof is in the pudding, and the major developments in our method are usually inspired by the feedback from chefs and home cooks whose knives we've sharpened, and by the brief time I spent in Japan learning from sharpeners I admire.

Teaching students from scratch in our sharpening classes forced me to figure out a way to break the process down into its essential components, identify what's truly important, and, in the end, question my own assumptions before sending the students out into the world. The technique here is the one I use and the one I teach.

Although the first sharpening orders I ever put out into the world were far below anything I'd be proud to pass through the shop today, they were miles ahead of a dull knife. For as much mastery as there is to be had in sharpening, I am a firm believer that plenty of low-hanging, delicious fruits—imperfectly sharpened but perfectly usable knives—are there for the picking. The only things you need to make a good edge are a simple technique that's easy to repeat, a moderate grasp of the steps, a few decent stones, and some persistence.

And so, I make no claim that the technique offered here is "the right way to do it" or superior to other techniques. But its simplicity makes it easy to replicate, and it should help you isolate what works and what doesn't work for each knife and edge. It serves us well to take a cue from our friend Auguste Escoffier here: *faites simple*—"make it simple."

THE SHARPENING PROCESS

I like to stress simplicity with my sharpening. The fewer moving targets you shoot for, both mentally and physically, the easier it will be to track your movements, stay consistent, and isolate and trace any edge flaws back to aspects of your technique (and there will always be flaws). In time, the number of flaws will diminish as the motions encode themselves into your muscle memory—as your hands and mind get more and more practice thinking about how certain knives will wear and how to sharpen them the best way possible.

Sharpening is about far more than learning one technique, however. Not all knives are sharpened the same way, and you'll need to understand a few things about edges if you want to know how a certain sharpening stone can make a knife come to life. Along the way, experimentation and actually using your knives to cut food are essential steps to learning what works. Your edge might cut through a sheet of paper like a dream, but if it doesn't work well on an onion, what's it good for? I'll try to tackle all the questions that helped me get started, in hopes you'll find a way to get started. If you've already clocked some time on a sharpening stone, maybe you'll pick up a few tips. Sharpening is an evolving practice, and this is less a treatise than a diving board into the part of the pool I would recommend swimming in first.

A Look at Dullards

The first step to sharpening is learning how to read a dull blade. To do that, you must look for two things: edge and geometry defects.

If I sharpen two identical edges the same way, they'll invariably come back to me in different states of wear and tear. One might be rounded to one side and the other nicked and chipped all over the middle of the blade. The way a knife wears down depends on the way it's used, on what kind of food, and on what type of cutting board. Plus, a single technique can yield different results depending on the degree of precision with which it's practiced.

Think of a sharp edge as an uninterrupted angle. In contrast, a dull edge is disrupted by rounding, breaking, or bending. This can happen either slowly over time or quickly when the steel is stressed and damaged through improper use. The main difference between a sharp knife and a dull knife is whether the bevels intersect in a crisp, clean fashion. As a knife dulls, it wears away that point of intersection.

Assessing the Edge

First, hold the profile of the blade against a contrasting background and look for chips and nicks. Then, look at the edge straight on to check for any bruised or rounded spots. Tilting it to catch the light can help. These spots will reveal themselves by reflecting a spot of light where the blade is bent or damaged. A roughed-up spot might look a little glittery in the light, and an evenly dulled rounded edge will shine right back at you. The thinnest, sharpest portions of the blade won't reflect any light.

Sharpness

Now test the sharpness of the edge. Many of us were taught that brushing our thumb across the knife edge, perpendicular to the blade, is a good way

to test it. A dull knife will slide smoothly across your skin, but a sharp knife will grab. If the edge is dull but bent to one side, this method can trick you into thinking the edge is actually sharp. You won't find that out until you try to cut something. I think it's better to start with the pads of your fingers on the side of the knife blade, and move them outward away from the edge. Repeat this motion as you inch toward the tip of the blade, until you've felt the whole edge on both sides.

If the knife is bent more to one side than the other, each side of the edge will feel different. The rounded side will almost pull your fingers across it, while the side the edge is curving toward will feel rougher and catch your fingertips. You can also feel for nicks and other isolated damage this way.

Blade Geometry and Thickness Behind the Edge

Two variables control how smoothly a knife cuts through food: the sharpness of the edge and the thickness of the blade from the center down to the edge. When I say "behind the edge," I'm talking in fractions of a millimeter—just before the edge drops off.

Consider the difference between a splitting maul and an ax. A maul, meant for splitting firewood, needs a blade that is wider and thicker than an ax, which is designed for felling trees or carving wood. To split wood, a maul needs to wedge it apart along the grain, which makes the actual edge less important than the shape and width of the body of the maul as it is forced through the wood. Axes, on the other end, rely on a sharp edge to penetrate wood against the grain. Even with a sharp edge, a maul is too thick to slide through wood against the grain. In the case of a knife's edge, you want more ax than maul.

A blade that's thick behind the cutting edge diverts force from the edge, regardless of whether it's sharp or not, and creates resistance that requires some force to push the blade through. That kind of force can bruise and damage the food you're slicing.

If a blade is sharp enough to shave a beard, cut paper, and cleave a length of silk floating down a stream but is thick behind the edge, it will work horribly on an onion.

Assessing Thickness

Testing the thickness behind the edge will help you determine how much to shave off the bevels of the knife to ensure it glides cleanly through food.

To test the thickness of an edge, pinch the body of the blade between your thumb and forefinger and draw those fingers outward toward the edge. Does the blade taper gradually and smoothly until you are pinching only air? Or does it end abruptly in a thick ledge? If the latter, the edge is on the thick side.

The Steel and the Edge

The type of steel in your knife has a lot to do with the type of edge it can take and how long it can hold it. Harder steels hold an edge longer than softer steels and can take a finer edge because the steel won't fold as easily if ground thinly. Softer steels generally won't take a fine edge or hold an edge as long as hard steels, but they can be tougher and more resistant to damage. Understanding the type of steel in your knife and how it will wear can help you build a sharpening strategy that ensures the knife works as effectively as possible.

Japanese knives come in different thickness, but they're usually made from harder steels than Western knives. Western knives, like a chef knife, are more often made of softer steels because they're generally designed for a larger variety of tasks and require a material durable and tough enough to withstand unpredictable usage. That means that your sharpening goals will depend on what the steel is capable of doing and how it will wear down. A Western knife will often need a slightly thicker edge than a Japanese knife, though not always. Sometimes a Western knife will be made as thinly as some Japanese knives, and more frequent sharpening is expected. In this case, you can try to put a fine edge on your knife but know that it will wear quickly.

THE STONES

Many different types of naturally occurring stone will sharpen a knife. Some stones, like sandstone, are friable—releasing their grit as they're worked—and others, like granite, whose abrasive crystals grow fused together, are not. From the Japanese perspective, the grit released from a friable stone is useful in sharpening. It mixes with water to flush away spent metal, keeping it from getting packed into the face of the stone, and it creates a lubricious mud or slurry, which can act as its own abrasive. Two things affect the way grit removes metal: the size of the abrasive particles and how deeply and firmly they are set into the stone mass. The more deeply they are set, the shallower the cuts into the steel. The shallower they are set, the deeper the cuts into the steel.

Japanese stones are divided into three classes of grit size: coarse, medium, and fine, or *arato*, *nakato*, and *awasedo*, respectively. Natural Japanese stones are typically sedimentary, but a few volcanic stones are in the medium, or *nakato*, range. Each of the grit sizes has a function, from removing significant amounts of metal and shaping, cleaning up the scratches from the coarser stones, and lightly removing metal on an edge that barely needs sharpening all the way to the shaping, final finishing, and polishing. Although you could use just one grit size, I recommend getting all three to bring a knife through a full course of sharpening.

The vast majority of Japanese sharpening stones used today are synthetic, modeled after the naturally occurring stones. Synthetic stones are more popular than natural ones because they're less expensive and easier to find, and coarse- or medium-grit synthetic stones work faster on newer types of stainless and high-alloy steels. A number of variables determine how a stone sharpens and what kind of an edge it leaves, including grit chemistry, consistency, density, and the type of binder. Some stones are slower to do large amounts of work but leave scratches that can be cleaned up easily; others work quickly but leave scratches that take longer to clean on the next stone. Always crucial, however, is the ratio of binder to grit.

Synthetic stones were originally made with a clay-like mixture of abrasive particles and binder, creating a brick that approximates the qualities of a sandstone or silt stone. The early stones are called "muddy" or "soft" stones because of how quickly they release their grit. These original formulas, as well as upgraded versions of them, are still used. As far as abrasives go, the widespread types are aluminum oxide, silicon carbide, and chromium oxide. These days, more modern techniques use different binders, many of which make harder stones sometimes called "ceramic stones," higher-quality versions of which will be denser in their ratio of abrasive to binder. A more abrasive stone has more abrasive particles available to do the cutting, which makes for a faster-cutting stone. The binder of the stone and the metal shavings of the knife combine to form a slurry that offers its own unique abrasive properties—properties that are often particularly important in sharpening traditional Japanese single-bevel knives. How firmly set into the binder those abrasive particles are, at what depth they're set, and what size they are controls how quickly a stone will cut different metals and different widths of bevel. For instance, one stone with dense but shallow cutting grit that forms a slurry readily might cut a wide, hard carbon-steel and iron bevel of a *deba* quickly, but cut a hard, wear-resistant stainless steel slowly.

Synthetic Japanese stones, like the naturals, are designed to be used with water, not oil (as some other stones are). They wear unevenly with use and

need to be manually flattened more often than harder "fused grit" stones, which don't come apart in the same way.

It is easy to lose sight of the forest for the trees here, but for beginners, the general rule of thumb is that if a stone is very cheap, it will be slow cutting. That said, expensive stones are not always proportionally better. It is far better to have all three grits of medium-quality stones than to have just a single high-quality stone, so use your budget accordingly. Coarser stones are usually less expensive than fine ones.

Harder stones wear more slowly and require less flattening, but they're harder to flatten when they need it. So, they're best attended to regularly. Softer stones generate a slurry faster, and in the case of fine, or *awasedo*, stone, less metal gets packed into their surface.

Natural Japanese Fine Whetstones

Natural stones with a medium and coarse grit tend to come from more commonly occuring rock, like sandstone, or the tuff harvested from Japan's famous volcanic-ash formations. Fine-grit stones, especially those with the right balance of softness and grit size for abrading steel, are the hardest to come by.

The stones we'll talk about here are the ones you've already met, the *awasedo* natural whetstones harvested from the Tamba terrain outside of Kyoto. Whetstones have been harvested from this band for eight hundred years and have played a crucial part in the development of Japanese edged tools through the extremely fine edge they could produce on a hard *tamahagane* steel.

 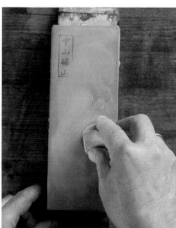 

Tennen toishi owe their power in part to their irregularity, softness, and friability. The grit on these stones is irregular—it is, after all, the compacted skeletons of asymmetrical creatures—and the sharpening happens by nature of the particles clumping and scratching more deeply than each individual particle. As these clumps are worked between the stone and blade, they get broken into smaller and smaller components, forming a mud that serves to polish as it scratches, forming a characteristically hazy finish and an irregularly scratched edge.

You might think inconsistency is bad when it comes to creating an edge, but this is the distinct advantage of a natural Japanese stone. The particles make an irregular sawtooth pattern on the edge, giving the edge more bite than a perfectly flush, toothless edge. The longer of those sawteeth will dull down to the next level of fresh abrasive teeth, which is how irregularity gives naturally finished edges their lasting power.

On the bevel of the blade, the irregular scratch pattern creates a hazy surface texture that releases food with more ease than a perfectly smooth, mirror-finish surface, which holds food by suction. This might not seem that consequential, but when you drop a single-bevel knife like a *yanagi* through a fish or a *usuba* through a vegetable, that hazy surface finish will lessen the pulling and tearing that breaks cells and causes the oxidation that you didn't know you needed to worry about until I started talking about it so much.

CHOOSING A GRIT SIZE

After the dull edge has been assessed, it's time to pick the appropriate abrasive to do the work. If you see chips or bright reflections of light off the edge itself, the blade needs a coarse, aggressive grit to remove enough metal to restore the geometry of the edge. If less damage is evident, a stone with a shallower grit will do. If it is a best-case scenario and the knife is only a little dull and no chipping or damage is discernible, a quick trip to the finishing stone should do the trick.

Coarse-grain particles cut deep paths through a knife bevel, leaving a large tooth on the edge. Conversely, fine-grain particles cut shallow grooves, leaving small teeth on the edge. These teeth interact with food as well as with the cutting board beneath it in different ways. A medium 1,000-grit scratch pattern tooth grabs and tears into food, especially when swept backward and forward in a slicing motion. Larger teeth, being long, unsupported structures, are more prone to deformation when they hit the cutting board and can bend or break on contact. The tiny teeth from a superfine 10,000-grit stone don't bite like the large teeth, and some of their power washes out when slicing, but they can wield a push cut more effectively. Small teeth hold up better to contact with a cutting board because they are better supported, and when they are bent or broken, the damaged area is smaller.

Given these features, you'd think that a 10,000-grit stone is the way to go. It has the bonus of good looks and a glamorous, mirror-like finish, but not all is as it seems.

What to Consider When Choosing a Grit Size

Knives are at their sharpest fresh off the stone, but they begin to wear immediately, so 90 percent of the time you spend with them, they'll be in some degree of sharpness less than that. That means that what matters most is how your knife performs in this "sharp enough" middle ground, and that depends on steel type and scratch pattern on its edge.

To start, consider the knife's purpose. A blade that needs to grab fibers, like a boning knife that moves through ligaments and tough connective tissue, benefits from a coarse, toothy edge. A chef knife that is used on onions will benefit from a finer edge because it needs to minimize its damage to the cells of the onion. A coarse edge, by contrast, would tear the cells as it cuts, leaving them prone to oxidation.

I don't typically use a grit size above 8,000 on most double-bevel knives because, and this is a matter of opinion, I think that past a certain fineness on a synthetic stone, the feel of cutting gets a little washed out. A superfine edge sometimes slides as it cuts, taking away from the "feedback" and feeling, and ironically, it's not as sharp. An effective edge gives feedback, or *kireaji*, that you won't find on the edge of most double-bevel knives finished on a superfine-grit synthetic stone. But we do keep a superfine-grit synthetic stone in our arsenal at Bernal Cutlery, and its job is typically to finish the *urasaki* single-bevel knives, providing a platform for the *kireaji* of an *omote* bevel finished fine with *tennen toishi* or coarser sometimes for *deba* to provide added tooth, but mostly it is used to pre-polish shaving razors before a fine, hard *tennen toishi*. Of course, many razor people would take issue with that approach as well. See how easy it is to step in something when sharpening?

Learning to sharpen should make you a fussbudget about edge quality, but it should not turn you into a chronically dissatisfied cook. You'll come to consider the "pop" as you start to cut through an onion, or the way your knife slides along a pepper skin, as the sign of a dull knife, even if that's what you've always been used to. I hope this inspires you to maintain your knives well and often.

But as you learn, remember that the confines of "sharp" exist on a wide and generous bell curve that narrows as you progress. Sharpening a knife well enough to sink, not pop, through the skin of an onion is only a few minutes and some elbow grease away.

The Four Basic Stages of Sharpening

1   Assess the degree of dullness and any possible repair work and or thinning the knife needs and then select your beginning stone.

2   Set the bevels (or create an intersection of the two sides of the knife).

3   Remove scratches from coarsest up to finish grit.

4   Remove the burr.

This is the most condensed version of sharpening; step 3 might need two or even three stones to accomplish, thinning the blade might come before step 2 for advanced sharpening, and step 5 would be to dull the knife by using it so it can be sharpened again, keeping you in practice.

SHARPENING A DOUBLE-BEVEL KNIFE

Works on: standard Western knife, *migaki* finished Japanese knife without a pronounced *kiriba* grind

Setting the Bevels

Begin by holding the knife in a pinch grip, with your forefinger and thumb pinching the spine just above the handle. Now move your thumb onto the heel of the knife, close to the edge, and move your forefinger to the top of the spine for knives that are wider than a paring or utility blade. Consider this knife, your hand, and your forearm as a single fused unit—as a straight line from elbow to wrist to knife tip, like the stars of Orion's Belt.

Hold the knife at roughly 45 degrees to the long centerline of the stone, and place the blade on the bottom edge of the stone closest to you. The edge should be facing you, and the heel of the knife should be on the stone. You will work from the heel to the tip.

Place the forefinger and middle finger of your other hand at the bottom of the knife's heel, near the edge and close to your other thumb. The angle between the knife blade and the stone is important here. Generally, 15 degrees is a good angle for sharpening thin knives, which you can approximate by stacking two U.S. quarters under the edge of the spine. The width of the knife will affect the angle created by two quarters, so for wider knives, tilt the spine up another hair or two.

Now, without moving your wrists, maintain the 45-degree angle of approach and the 15-degree blade angle and move the knife in a straight line back and forth along the stone. Keep a light pressure consistent on the forward and back strokes, moving the knife along an imaginary (or, if you have a pencil, real) center axis, right down the middle of the stone. Apply just enough pressure with the fingers on your secondary hand to make the tips of your fingers go white, but no more. The job of these fingers is to maintain and focus contact with the stone and blade, not to grind the blade into the stone. The stone will do the work if the correct amount of contact is made and if there is enough motion.

SAFETY NOTE: I encourage you to use the full length of the sharpening stone, but that can leave a gap between the blade and the end of the stone on the upstroke, creating a little guillotine primed for catching a finger. **DO NOT** allow your stabilizing fingers to extend past the end of the stone. The word *guillotine* should frighten you enough, but just in case it doesn't, let me say it again: guillotine. Consider the knife a life raft for your fingers, and stay on board.

Continue this basic movement until you can maintain some consistency in the angle, light pressure, and direction of the knife, all the while keeping the brace from your elbow to wrist intact. Once you're comfortable, start moving your secondary fingers in tiny increments up the blade in the direction of the tip, pulling the knife handle away from the stone toward your elbow as you go just enough to keep those fingers in the center of the stone. Keep on pacing back and forth.

Continue as before, until you reach the place where the edge curves up to form the tip of the knife. Now flip the knife over and hold it in your other hand. We'll come back to the tip after working the other side.

To sharpen the opposing side, hold the knife in the same modified pinch grip as before, keeping your elbow, wrist, and the knife tip in a straight line, and repeat what you did on the first side. If this new grip is with your non-dominant hand, you may notice that the supporting fingers of your dominant hand, now pressed against the blade, might be driving more of the motion. Great! Use them to keep your rhythm and angles consistent.

Once you reach the place where the blade curves to reach the tip, stop. The knife tip is more challenging to sharpen, and keeping the angles of the bevels consistent with the rest of the knife that you've sharpened up to here depends on keeping the angle of the knife steady. You'll be tempted to twist your wrist to change the angle, but all you have to do is lift the handle along with your foream and elbow to rock the edge into contact with the stone. Resist the temptation to twist the knife here, as it will ruin the angle of the edge at the tip, making it too thin or too thick.

Now, if all went as planned, the bevels should be ground in just enough to have eliminated any tiny chips or nicks in the edge. Once the two bevels intersect, they will leave a slight overhang of metal called a burr. You can check for this intersection by feeling outward with your fingers toward and past the edge, looking for a fringe-like overhang. Depending on the kind of steel, a burr can be a hard, rough edge; a stiff, brush-like edge resembling

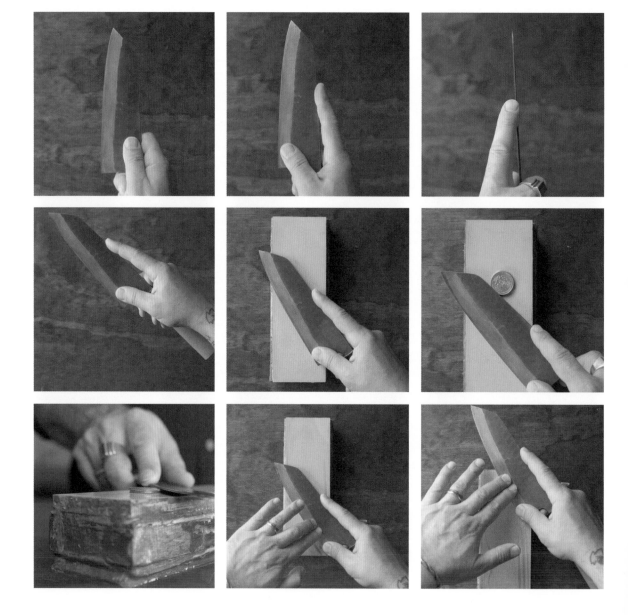

tinsel; or a little of both. It's easy to confuse the feeling of a burr with a bent edge, but a burr will shift from side to side with sharpening and can usually be moved with a fingernail.

Removing Scratches

Now that you've set the bevels, it's time to remove the coarse- or medium-grit scratches just laid down by shifting to a stone with a finer grit. If you began with a coarse stone, move to a medium stone. If you began with a medium-grit stone, move to a fine-grit finishing stone. Because you are working with narrow bevels here, you can skip some grit sizes and move from a coarse to a fine stone. But if you were to try this with wide bevels, don't skip from an ultra-coarse grit like 220 to something as fine as 1,000. If possible, use an intermediary 400 to 600 grit to help take out the large scratches. Otherwise, you'll be making a lot more work for yourself trying to remove more metal than a fine-grit stone is able to do quickly. Likewise, going from a 1,000-grit to an 8,000-grit stone will be slow going on a wide bevel. So, don't skip grit sizes if you are working on wide bevels.

Eliminating scratches takes less time than setting the bevels, so this part should typically move more quickly. Be thorough, however, allowing for a few minutes on each side.

Now switch to a finishing stone and repeat the process again, removing the burr before you're done.

Once you've finished sharpening, pull the edge along a piece of leather (see photos a-d on page 134), a rubberized cork with a fine abrasive (my favorite material; I like 1 micron chromium oxide), or even felt, newspaper, or cardboard. This is the sharpest your knife will ever be until you sharpen it next.

AVOID AWKWARD MOTIONS

In the beginning, you'll probably find yourself straying from the 45-degree position of the knife by bending your wrist and arcing the line of your elbow to knife tip. If you notice this, correct it gently until it is second nature. You might also be prone to a scooping motion: typically a steeper angle on the way back toward you. This is a bad thing to do, so don't do it. It creates greater irregularity in your technique when you start, and it takes years of practice to build the muscle memory and consistency. Along the way, it's all about knowing what to look for. Because tracking the angle of your blade or your patterns can be tricky, either consider asking a friend to film your technique or set up a tripod.

Thinning a Knife

Each time a knife is sharpened and it loses metal from its edge, the whole blade gets a little bit narrower. Its shape also becomes more wedge-like as the edge creeps close to the center of the blade, where it's thicker. On a hand-ground or thick blade with a wide bevel, or *kiriba*, it is a good idea to compensate for the metal you take off the edge by also shaving some off the sides directly behind the edge. Although thinning a knife provides a huge benefit by the same ax-versus-maul logic that explains a good edge, some people insist it destroys the finish of the knife and adds flaws. While it is true that it changes the appearance of the blade's finish, it is also a purely aesthetic perspective, and if you plan to use your knife, I urge you to get over that concern. If you plan only to look at it, stop here, invest in a high-quality spoon, and stick to foods that don't require cutting, like yogurt.

HOW TO KNOW IF YOUR KNIFE NEEDS THINNING

As soon as you're comfortable with the basic sharpening technique, you can try your hand at thinning a knife.

Make sure your stone is perfectly level. Place the area to be thinned directly on the stone in the same position you used for working the edge: edge on the stone and spine tilted up, angled at 15 degrees. With a light but decisive pressure, scrub the face on the stone in a back-and-forth motion. Overdoing the pressure won't speed up the sharpening. It will only give you less control. If you have a knife with this sort of grind, I highly recommend thinning as a regular practice.

Testing a Sharpened Blade

To check for sharpness and edge geometry, you can either feel your way around the knife by the same method you used to assess the blade or you can run one of the following flashy tests: Hold a piece of paper vertically and try

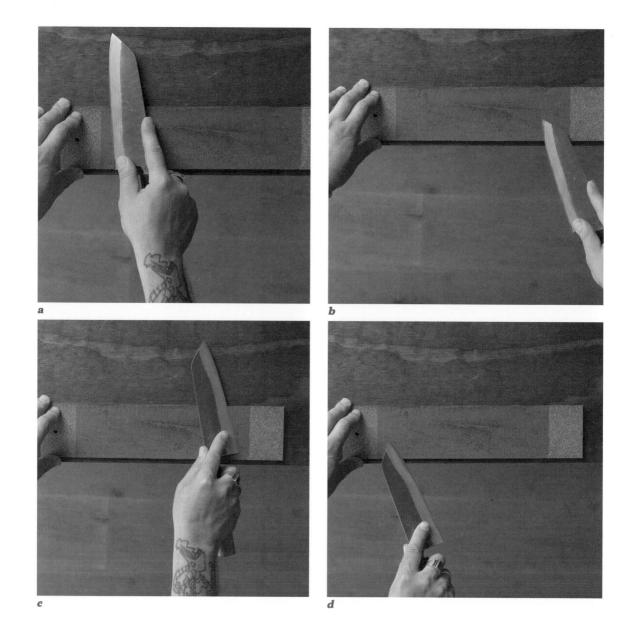

to slice through it top to bottom. Gently—and I mean really, really gently—slide the edge against your arm hair at a shallow angle without touching your skin. A sharp blade will cut, and anything duller will pull at your hair. At the shop, we think the best test is the material that the knife is meant for, and so because none of us eats paper or hair—at least intentionally—we keep a handful of carrots behind the counter.

Carrots are dense and sensitive to wedging, which means they'll give feedback about edge geometry, and their surfaces offer insight into the quality of an edge. Plus, they're way less messy than tomatoes, and it takes only a few quick cuts to assess the shape and toothiness of an edge.

Slice a few thin rounds off the end of a carrot to measure whether the bevels are flat or rounded. A V-shaped edge will tend to cut straight down, and a more rounded U-shaped edge will guide the cut in a semicircle, producing a rounded cut. A lengthwise cut along a piece of carrot 3 to 4 in/ 7.5 to 10 cm long will give a good indication of the wedging that a blade makes: a blade thick behind its bevels will wedge and split the carrot, and a thin blade will slip through easily.

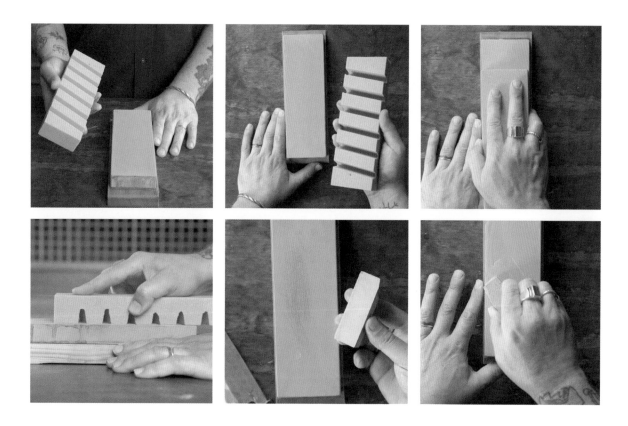

SHARPENING A SINGLE-BEVEL KNIFE

Works on: *deba, usuba, yanagi*

Sharpening Japanese-style single-bevel knives requires a few different techniques from double-bevel knives. Generally, they're a bit more work, and occasionally a lot more if they have been poorly sharpened and need to be corrected. But a single bevel also makes a knife easier to sharpen in at least one way: it eliminates the need to guess at angles, which can be a source of frustration for novice sharpeners. Single-bevel knives instruct the sharpener where to begin because the angle is already ground into them. Good technique is needed to preserve this geometry and to make adjustments for the user's experience level (a rounder, thicker edge is better for a less-experienced user, who will be quicker to damage it).

When a single-bevel knife is well made, there are subtle shifts in the angle of the bevel as you travel from the heel to the tip. Typically, the angle is steeper at the heel and shallower toward the tip. The difference is more pronounced in some knives like the *deba*, where the work done at the tip and belly (delicate filleting) is much different from the work performed at the heel (bone breaking).

The general consensus is that a thicker edge that chips less is desirable for beginning knife users, and as a user becomes more experienced, a thinner edge is more acceptable.

First, let's consider the shape and parts of the single-bevel knife. This is not an exhaustive list of what makes up a single-bevel knife; these are the elements whose Japanese names have no English or other Western equivalent.

OMOTE: The outer side of the blade that the single bevel is on, or the knuckle side when you hold the knife. The knuckle side of a double-bevel knife is also called the *omote*.

URA: The inner side of the blade opposite the single bevel, or the palm side when you hold the knife. The surface is ground concave. The palm side of a double-bevel knife is also called an *ura*.

KIRIBA: This is another name for the bevel, or blade path. If a double-bevel knife has wide primary bevels, they are called *kiriba* as well.

SHINOGI: The boundary between the *kiriba* and the face, or flat, of the *omote*.

URAOSHI: This is the thin, flat bevel on the concave-ground *ura*. The *uraoshi* is formed by placing the *ura* flat on a stone. The edges will be the only place of contact, and the *uraoshi* will be formed then.

JIGANE: The soft cladding of a knife.

HAGANE: The hard steel at the cutting edge.

Getting Started

Begin by assessing the blade's condition. There are different ways that different steels will dull, but most of the time, single-bevel knives get chipped before they get rounded. Check for damage as you would with a double-bevel knife (see page 129–130). If there are significant chips or very heavy rounding, you will need to begin with a coarse stone (220 to 600 grit). If the dulling is less dramatic and no chips are easily discernible, you can start with a medium-grit stone (800 to 1,200 grit).

Once you have established how much work is needed (with a single bevel, it's always a little more than it seems) and you have selected your starting stone, you must next master the correct grip. Rather than holding the handle with the tip in line with the forearm, like you did on the double bevel, you need to hold the knife slightly sideways on the stone. With the spine and corner of the stone at a 90-degree (right) angle, bring the knife in contact with the stone at the heel. Start with the thumb, index, and middle finger of your secondary hand all over the *shinogi* line. Start with moderate pressure and try not to jam the knife into the stone. Remember, the motion, not the pressure, does the work.

Keeping the thumb and secondary fingers over the stone (don't let them trail off over the corner), move straight back and forth over the length of the stone. Long strokes are preferable to short ones, but it takes some practice to work up to them.

> **SAFETY NOTE:** Before you go any further, return to the safety note on pages 128–129 and read it carefully. The same guillotine warning applies when sharpening single-bevel knives as when sharpening their double-bevel kin.

Keeping the point of contact at the *shinogi*, start to walk your fingers up the blade toward the tip, keeping the fingers over the stone as you go. By not pushing too hard, you diminish the risk of rounding over the *shinogi* between the top of the *kiriba* and the face of the blade. There is still no contact with the edge and that's fine. At this point, you are working this area to maintain the geometry of the *kiriba*. Once you get near the tip, start to lift the handle a bit to help the *kiriba* and *shinogi* stay consistent to the original grind. As you work, feel for the shape of the *kiriba* under your fingers.

Now move your thumb and secondary fingers to roughly the center of the *kiriba*, over the hazy *kasumi* line (the boundary between the soft *jigane* and the hard *hagane*). The blade will slide more easily over the stone, but the hard steel will be removed a little more slowly. Work your way up to the tip again, remembering to lift the handle a little (but don't twist it!) at the tip.

The next, or third, position will vary in technique depending on the skill level of the knife user. A highly experienced user might cut with straighter angles and will rarely exceed appropriate pressure, while a less-experienced user tends to twist or wobble during cutting and often uses too much pressure, damaging the edge. When sharpening for anyone but a very experienced user, a bit of *hamaguri*, or clamshell shape, is useful to prevent chipping. If working on a *deba*, the heel generally gets a *hamaguri* to help toughen it up for work like breaking fish bones. To put in a little clamshell shape with a slightly convex edge, lift the spine slightly during position three, where the fingers are closer to the edge.

When you feel a burr on the *ura* side (which is yet to be sharpened), the work with this first stone is finished. There will likely be a few chips to smooth out, and this must be done in the areas all around the chip, not just right over it.

If you are sharpening a *usuba* or other totally flat-edged knife, I recommend removing chips by working just the edge at a very steep angle. This is one of the more difficult jobs in sharpening single-bevel knives. With a *usuba*, however, it's especially important to keep its flat edge perfectly flat for working on a cutting board, and if that edge is chipped, it takes a lot more foresight not to create a wavy edge as you file it down. But no matter the type of knife, it is important to preserve the overall shape of the edge and to work the entire edge, and not only random areas.

Once this work has been accomplished, go to the next stone. If you have been working with a coarse stone, use the same technique with a medium stone, and if you started with a medium stone, move to a fine or medium-fine stone, depending on the knife and the stones you have.

Once you've worked through a medium-grit stone, regardless of whether it's the first or second stone, examine the ura side and look at the uraoshi: is it very, very narrow near the edge—less than 1 mm/$\frac{3}{64}$ in? If metal was not initially removed from the edge, and the uraoshi is still intact, do the steps below using a fine stone on the uraoshi. But if you began with a large chip and removed some of the edge on the bevel side, effectively removing the uraoshi too, you may need to reestablish it with a medium stone before moving to a fine stone to finish it. It is important to resist flattening the ura too much by overgrinding the uraoshi. The concavity of the ura is what gives single-bevel knives their smooth cutting feel.

To reestablish the uraoshi on a medium stone, you can sharpen the uraoshi like this: Without switching hands (like you did to work on the opposite side of the double-bevel knife), lay the blade dead flat on the stone. With light pressure near the edge, slide the knife back and forth. Keep a gentle pressure near the edge, and don't let the blade get overworked near the spine. Remember that the spine area of the uraoshi is soft jigane (rather than the hard hagane near the edge), which will shed metal easily. Do a little work—a few minutes' worth—and examine the uraoshi. You're not shooting to make it too wide here. Opinions differ on how much the uraoshi should be sharpened. Some recommend doing it more in order to strengthen the edge, while others caution against it because over time it can lead to a flattened ura. I think the uraoshi should be roughly 0.5 to 3 mm/$\frac{1}{64}$ to scant $\frac{1}{8}$ in wide; the thinner the edge is, the more likely it will chip, but the thicker it is, the more it will drag. In cases of very little or no uraoshi, the edge will fail easily, but with a very thick uraoshi, the ura will stick and pull against whatever you are cutting. Once you've reestablished the uraoshi, or if the uraoshi was kept intact, repeat these steps with a fine stone.

Next, remove the scratches on the *kiriba* using the same technique you used with the previous stone(s).

Most knives benefit from a tiny secondary bevel, known as an *itoba*, or microbevel, at the very edge of the *omote*. An *itoba* will strengthen the edge and will help to make the very end of the edge more chip resistant. If you've already rounded the bevel with a significant *hamaguri*, I would caution against putting in too much *itoba*. But if your *kiriba* is very flat, an *itoba* is strongly recommended. Some sharpeners discover the *itoba* for the first time and think they've found the magic bullet for sharpening single-bevel knives. You could take this shortcut once or twice on a barely dull edge with a very fine stone before you work your way back up to the *kiriba*, but if you do that too much, you'll dramatically increase the work you must do on the *kiriba*. Creating an *itoba* on the *uraoshi* is a bad idea and creates more work on the *kiriba* side to restore the original geometry. So stick to the *omote*.

To create a microbevel, simply lift the angle of the *omote* side high and give it two or three passes. Finish off by going back to the *ura* for a minute, being careful to stay flat on the *uraoshi* rather than angled up as on the *omote*.

A Word on Finishing Grits

There are different approaches to finishing an edge. It's not uncommon to put a new, semirough edge on a knife every day, especially the knives in heavy rotation in a professional kitchen. It will keep the edge "sharp enough" to use but will use up the knife faster. Most cooks prefer an edge that's marginally finer than the typical 1,000-grit grade of a medium-grit stone. Some insist on finishing the edge as finely as humanly possible, but I think that's a bad idea. To choose a finishing grit, ask yourself, what teeth are best for the job the knife will be doing? Experiment with different grits.

Chapter 5

Methods and Recipes

Having good knives is only as good as having the opportunities to use them, and it would be unnatural to end this book anywhere but the kitchen.

Our shop sits between knife makers and chefs. There is an indirect but distinct line of communication between them, and we get to be a kind of transmission tower that sends and receives feedback in both directions. My growth as a sharpener and knife dealer has been greatly informed by my own experience—I'm able to think with my hands and teach myself—but none of my real growth would have happened were it not for the people with whom I have had a chance to collaborate. Whenever I am able to turn someone, a home or professional cook, on to a new knife and get his or her feedback, I'm able to flesh out a little bit more of that knife or of the character of its edge.

There were many chefs I wanted to include here, but with a limited amount of space, it was important to get a mix of styles and stories. I have done my best to provide a good cross section of the cooks who are regulars in the shop and whose perceptions have taught all of us who work there more about our sharpening and about the knives we carry than we ever could have learned on our own. In the following pages, you'll see traditional cutting techniques, and then those same rules getting smashed to pieces by another chef. I hope that by including a number of styles here, you will add a few more skills to your quiver.

Knife Skills and Strategy with Bernal Cutlery's Sam Rezendes

Sam was a regular at our shop in Bernal Heights before we moved to the Mission District. Back then, he was cooking at State Bird Provisions, and I recruited him to teach knife skills from our shop. Eventually he became a full-fledged member of our shop family in 2013. He trained in both Western and Japanese cooking and brings that sensibility to his teaching.

The first thing I ask myself when I'm deciding how to cut something is, how will the heat be applied? Will I be grilling? Roasting low and slow? Blanching, stewing, or simmering? Larger pieces take longer to break down, so I'll go bigger for a long roast. If I'm cooking in hot oil, I might want pieces that will cook all the way through before the outsides burn, in which case I'll make them small. What I do know is that evenly sized pieces will cook at the same rate, and usually that's a good thing.

Cutting changes flavor, too. Thinly cut celery requires less chewing and has more surface area than a thicker cut, which means more oxidation before it hits your tongue. Thinner, delicate celery pieces are sweet and tender. On the flip side, thick cuts need to be chewed a little more, which changes the way you contact their flavor. As you chew, you'll notice the vegetal flavor of the juice and the texture of the fibers. Chopped garlic can taste more pungent than thinly sliced garlic. A radish cut into chunks will be spicy, while the same radish thinly sliced will be sweet. Cured meats act this way, too. Try chowing down on a chunk of salty salami after tasting a thin slice of the same meat.

As you strategize how you'll cut something, ask yourself: "Do I want to incorporate the flavor of the vegetable, or do I want it to stand on its own?" In the case of a potpie, a nice big chunk of onion can be delicious. In a salsa, you probably want it to blend in. When you start to think about it this way, knife skills become a kind of subtle flavor control. Take the same recipe and cut the onion two different ways, and they will taste different, for better or worse.

And then there's the question of evenness. Uniformity is good when we want things to cook evenly, but what if we don't? The oblique, or irregular, cut is all about imperfection. When ingredients are cut into different shapes or sizes, they'll cook, marinate, and cure unevenly, which adds more layers of texture and flavor to a dish than if everything were cut the same. It's a functional tool, but the best part of the oblique cut is that it cracks apart the idea that your knife work must be perfect. Traditional techniques like to focus on perfection, but besides bruising your ego, rigid rules can cause some problems, too. If 90 percent of your carrots are uniformly cut, those last few outliers will stand out like sore orange thumbs. If they're all a little irregular, they'll look like a beautiful, natural nest of food. Be consistent with your inconsistency and don't sweat the right angles too much.

Simply put, there are no rules. Even if there were, they probably wouldn't work every time. Vegetables vary in moisture, starch, and sugar content throughout the harvest and into cold storage, and every animal you butcher will be different. You are the chef, and although a good recipe is a wonderful guide, there is no substitute for mindful tasting and a little trial and error. So cut, taste, move on, and don't try to be too perfect.

The techniques here represent the classes of techniques at the core of a cook's skill set, and every cook has his or her own cutting style within these classes. Here I have attempted to break down the different *kinds* of cuts, not so much the cuts themselves, in hopes of empowering you to improvise. You'll find variations of all these across the following recipes, and you're welcome to modify them.

Before you cut, it's useful to understand your knife as a saw: a long edge with tiny teeth. Those teeth will naturally engage with whatever you cut and will probably require less pressure to do the work than you might expect. Imagine the difference between pressing and sliding your finger pad along a knife blade. Even a sharp knife might not cut you if you press your finger to it, but if you *slide* your finger along the edge, you can bet it'll slip through the skin and draw some blood. In the same way, pushing a band saw against a log won't do much, as it's the motion that matters. So in the case of cutting, certain techniques might look as if a knife is pressing through a carrot, but some sliding movement is usually doing most of the work.

As far as choosing the right knife for a job, shorter knives offer more control at the tip, and the length of a longer knife provides more surface area for slicing. The latter is also better for executing the rocking push cut, which you'll find to be one of the more useful styles here.

USING THE PINCH GRIP

Holding your knife with a pinch grip usually allows for more precision and control than just grasping the handle. To use the pinch grip, pinch the blade above the handle between your thumb and the knuckles of your index finger and bring the rest of your hand around the handle. Resist the urge to put your index finger on the spine.

For larger jobs, like halving a cabbage or winter squash, take your fingers off the blade and wrap them around the handle for a firmer grip and more clearance. Whenever possible, however, use the pinch grip for more control.

THE PUSH CUTS

These are the foundational moves of the cutting repertoire of the chef knife. Push cuts are versatile and, by nature of the forward-moving motion and the ergonomics of your body behind it, generate more pressure with less exertion from your wrist than a pull cut. Because the blade is sliding rather than pressing down through whatever it cuts, the teeth of the edge do much of the work that would otherwise fall to your muscles. One caveat: Driving a fine, sharp edge into the board repeatedly as you do here will dull or even damage it. If you're used to the way dull knives cut, you might overwork a sharp knife by punching it against the board with unnecessary force. So all you rookies out there who are new to working with a sharp edge or a thin knife need to ease up.

THE CLASSIC PUSH CUT

To cut something large, like a cabbage, lift the blade of the knife off the cutting board and hold it level. Move the knife from tip to heel, pushing forward more than pressing down as you slide the edge against whatever you're cutting. Lift the handle as you pull it back toward you, then let it rest against the food as you push it again. A sharp edge will do the work. When cutting something hard or resistant, commit to the work and punch the knife through. Don't use a half step or too little pressure. If your first cut doesn't make it all the way through, back out a little from where the blade stopped and then punch again.

THE ROCKING PUSH CUT

The rocking push cut, which is done with a curved blade, is best for smaller tasks. That's because the tip of the blade is left on the board for this cut, which means less clearance under the knife than for other techniques that call for lifting the blade off the board entirely. To execute this cut, place the tip of the knife on the board and bring the edge down as you push the knife forward, allowing the tip and the curved edge behind it to slide back and forth as you work. If you don't allow the tip to move, the blade will crush as it cuts. That will work, of course, but the quality of the cut is cleaner if you let it slide. This technique is especially effective in combination with your secondary hand, with the fingers used in a claw grip to hold the food and the knuckles braced against the blade to control thickness and to protect the fingers from an errant stroke. The rocking push cut is excellent for all of classic Western cuts: dice, julienne, chiffonade, and fine cut.

THE PULL CUTS

By nature of the motion, a pull cut diverts pressure away from the cutting junction. As you cut, that pressure gets distributed along the length of the edge's contact with the food. That's why a longer blade is typically the tool of choice here, as it allows you to apply more pressure—or rather, the same pressure over a longer period of time—with each stroke. The edge of a slicer is thinner at the heel than it is on other knives, which means it will engage with the food more quickly than a knife with a thicker heel, like a chef knife. The tip of a knife is also useful for pull cuts because it can follow contours easily, for example, splitting a celery stalk along its length. The distinction between pull cuts and slicing cuts isn't hard-and-fast, but to be pedantic, you can say that slicing cuts are pull cuts, but not all pull cuts are considered "slicing." For example, dragging the tip of a paring knife down the center of a leek would be a pull cut, but it would not be slicing.

THE SLICE

Generally speaking, "slicing" is a pulling stroke from the heel to the tip, usually done with a long blade. Slicing with a narrower blade creates less friction because it has less surface area than a wider blade, which makes it easier to move. That said, the heft of a wider blade will steer the edge a little straighter, making thinner cuts easier to do. To execute this cut, rest the heel of your knife against the item to be cut and pull back, drawing the length of the edge to its tip through the food. Slicing is excellent for separating cuts of cooked or uncooked meat and for removing cartilage or fat. In butchery, you'll accompany slicing by pulling the cut piece away from the main mass of meat, keeping the cutting juncture exposed.

THE PULLING SCORE CUT

A pulling score cut is ideal for precise work that follows the innate shape or curve of something because it naturally leads you to contours of the food fibers. It's also good for longer, deeper cuts. Consider splitting a broad, flat pork belly down the middle, or cutting a celery stalk down its length. A slicer's narrow tip comes in handy here, but you can also work with the tip of a chef knife. A shorter blade offers a bit more control, making the petty knife a good contender, too. To execute this cut, begin with the tip of the knife in contact with the board, just ahead of the food you are cutting. Drag the tip toward the food until the belly of the knife makes contact, and then continue to drag, keeping the point of the knife connected to the board as the belly behind the tip does the cutting.

THE PARING CUT

The paring, or peeling, cut is done in hand. The thumb and forefinger on the knife work together to pinch and slice at the same time, while the secondary hand helps to move the fruit or vegetable into the next position. In some cases, peelers have taken the place that paring knives once occupied, especially with potatoes, but many jobs remain that cannot be done with a peeler, like paring an artichoke or a spiny fruit. These tasks instead require a small, nimble knife, and the natural choice is a paring knife.

ON CUTTING AFTER COOKING

At the shop, our lessons focus on knife skills for cutting raw foods, but cutting foods after cooking is an important technique, too. Consider roast beef: cutting it before cooking makes for a far different dish than if it were cut after. This is the case with vegetables and fruits, too. Roasting large pieces of winter squash before slicing and then sautéing them yields a mix of different caramelized flavors and textures—the interior stays juicy and becomes sweeter, while the edges and broad face of the slices caramelize and crisp up in the pan. If you roast apples and then slice them, you end up with a mix of fresh, acidic apple flavor and sweet, caramelized edges. As you cook larger pieces of vegetables, think about cutting them afterward to expose more variation in flavor.

Cutting Onions with Armando "Tiny" Maes

Knife: chef knife, *gyuto*

Executive Chef // Fatima Restaurant and Inn, San Miguel de Allende, Guanajuato, Mexico

I got my first job when I was sixteen, stocking the salad bar at a Sizzler in Sacramento. I bought my first Japanese knife, the same one I'm using here, in 1998 when I worked at Rose Pistola, in North Beach in San Francisco. The tip broke in 2007, and I had it repaired, and these days it usually stays safe at home.

To be proficient in the kitchen, you have to know your way around an onion. It's the first thing I put in front of every cook I train. In my life, I've probably cut about 45,000 lb/20,400 kg of onions, and the three-part fine-slicing technique described here is what I use at home and what I teach my students. Cutting an onion this way exposes as much of the onion to the pan as possible, which is how you get the most flavor out of it. I admit that technically it's not the correct way to cut it, and you won't learn this in culinary school, but it works. An onion cut thinly is sweeter and won't offend the palate like a big chunk of raw onion will.

The fine-dicing and julienne techniques detailed here are also unorthodox. Some people say you shouldn't cut the core in, but I say to hell with that because I hate waste. This technique is safe, fast, and versatile, and it gives you a nice, even dice or julienne. The size of the dice depends on how thick you make your slices at each stage. To julienne onions, make the third stage of slices narrower than the first two.

Three-part Finely Sliced Onions

Lay the onion on its side and cut off both ends. Using a push stroke from tip to heel, cut the onion in half from end to end. Peel off the skin and lay each half cut side down. Using the push-cut stroke again, cut along the grain to divide each onion half into thirds.

Turn an onion piece 90 degrees, so the grain runs parallel to you. Bracing the knuckles of your secondary hand against the back of the blade as a guide, and starting at one end, move against the grain this time, cutting thin slices. Once you reach the opposite end, turn what remains of the onion flat so you can continue cutting safely. These last cuts will be uneven, so I usually throw them in the stockpot or scrap pile for using later.

Finely Diced or Julienned Onions

Lay the onion on its side and cut off both ends. Using a push stroke from tip to heel, cut the onion in half from end to end, then peel off the skin. Stand an onion half on its end, root side up, and use a push stroke to cut it into thin, vertical sheets, parallel to the flat face of the onion and working toward the rounded side, leaving a little connected at the bottom.

When you get to the end, turn the onion half onto its flat side and, using the tip of the knife, cut thin slices along the grain, piercing the onion at the connected end (be careful to keep it connected) and drawing the knife tip toward the flat, cut end. For a fine dice, make these slices thin. For julienne, make them a tad thicker.

To slice off the dice—or the julienne—turn the onion half so you will be cutting against the grain and use a push stroke from heel to tip, using the knuckles of your secondary hand to guide the knife and control the size of the dice or julienne.

Cut without Crying

1 Use a sharp knife.
2 Do not cut off the ends of the onion, and once you cut the onion in half, rinse the halves with cool water.
3 Put a piece of bread in your mouth to help filter the fumes as you breathe.

Thinly Sliced Onions

Lay the onion on its side and cut off both ends. Using a push stroke from tip to heel, cut the onion in half from end to end, then peel off the skin. Turn an onion half onto its side, with the root end closest to you. Push the knife tip into the end where the onion is still connected and, leaving that segment connected and working from either side, cut thin, vertical slices. Turn the onion half 90 degrees, and then using your secondary hand as a guide behind the knife, slice against the grain, cutting all the way to the end.

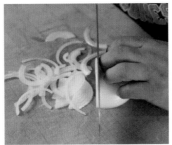

Melissa Perello's Autumn Squash Salad with Spiced Honey Vinaigrette

Knife: *gyuto*

Chef and Owner // Octavia

My first knife was a Victorinox 10-in/25-cm serrated chef knife, which I used at my first job as a prep cook when I was fifteen years old. I prepped a lot of burgers with that knife. These days, I use a thin carbon-steel chef knife for just about everything. My techniques have come about mostly out of necessity, and as a means of not wasting. That's why I use the oblique cut—an irregular cut means less waste, and I hate waste.

In this salad, the apples are cut both before and after they're cooked to mix the texture of a clean fresh cut and a cooked cut, and you can do the same with the squash and celery root. The pickled pepper slices are prepared a few hours ahead of serving and the fennel at least 30 minutes ahead, but I recommend letting them both marinate overnight, if possible. The vinaigrette recipe will make more than you need, but it will keep for up to a week in the refrigerator. Use it on everything.

The Cuts

To slice the pepper, with the pepper positioned horizontally, place the center of the blade on the pepper and pull from blade to tip to release slices about ½ inch/12 mm wide. Remove and discard the seeds.

To cut the squash, slice off the top to create a flat surface to stabilize it on the board. Hold the squash firmly with your secondary hand, keeping that arm close to your body. Starting with the tip of the knife on the squash, execute a push cut to halve it. Quarter each half lengthwise, or slice each half into 1½-in/4-cm wedges or strips. Now cut the quarters or strips into oblique, wedge-shaped cuts. These will look like small triangles.

To cut the apple, make three straight cuts from the top to the bottom of the apple, cutting the flesh away from the core. You should end up with a triangular-shaped core, which you discard. Cut each piece crosswise into small wedges.

To cut the celery root, use a vegetable peeler or a paring knife to cut away the greenish skin. Switch to the larger *gyuto* and cut away the roots. Cut the peeled celery root into oblique thick wedges against the grain, from side to side rather than from root to stem. Immerse the pieces in water with a few drops of fresh lemon juice or mild vinegar added to keep them from browning.

To slice the fennel, trim off any stalks and fronds (reserve for another use), then cut off a small, flat section from the side of the bulb to stabilize it on the board. Start with the tip end of the knife and push down, moving along the bulb as you cut slices ¼ in/6 mm thick.

Spiced Honey Vinaigrette

1½ Tbsp caraway seeds

1½ Tbsp coriander seeds

1½ Tbsp fennel pollen or fennel seeds

½ tsp black peppercorns

3 Tbsp wildflower honey

½ cup/120 ml sherry vinegar

½ cup/120 ml champagne vinegar

1½ tsp sea salt

1½ tsp whole-grain mustard or Colman's mustard powder

1 cup/240 ml extra-virgin olive oil

In a small pan over low heat, toast the caraway, coriander, and the fennel seeds, if using instead of fennel pollen. Transfer to a spice grinder or bladed coffee grinder, add the peppercorns, and pulverize to a fine powder. If using fennel pollen instead of fennel seed, add it to the ground spices.

In a blender, combine the ground spices, honey, vinegars, salt, and mustard. With the blender running, slowly add the oil until all of it has been incorporated, then continue to blend for 1 minute longer. The vinaigrette can be used right away, or it can be stored in an airtight container in the refrigerator for up to 1 week.

Pickled Pepper

1 fresh red sweet pepper (such as Jimmy Nardello, Italian frying, Gypsy, poblano, or, if you prefer a spicy condiment, Fresno), sliced and seeded (see The Cuts, page 159)

1 cup/240 ml sherry vinegar

½ cup/100 g sugar

1 Tbsp kosher salt

Place the pepper slices in a small heatproof bowl. In a small saucepan over low to medium heat, warm the vinegar to a low simmer, then add the sugar and salt and stir until dissolved. Pour the vinegar mixture over the pepper slices, let cool, cover, and let rest at room temperature for at least 3 hours or up to overnight.

Squash Salad

1 small red Kuri or kabocha squash,
about 2 lb/900 g, quartered and oblique
cut (see The Cuts, page 159)

Olive oil for sautéing, poaching, and
dressing

Sea salt and freshly cracked black
pepper

1 firm, tart apple (such as pippin,
Black Jonathan, Sierra Beauty, or
Gravenstein), oblique cut (see The Cuts,
page 160)

1 small celery root, about 1 lb/455 g,
oblique cut (see The Cuts, page 160)

Juice of 1 lemon

1 small fennel bulb, oblique cut (see
The Cuts, page 160)

1 small head frisée, leaves separated

1 bunch spigarello broccoli or lacinato
kale, stems and tough ribs removed and
leaves torn into 1-in/2.5-cm pieces

6 ripe figs, if in season, stemmed and
halved lengthwise, optional

Preheat the oven to 375°F/190°C.

Place the squash pieces in the center of a large sheet pan, drizzle with a couple of tablespoons of oil, season with salt and pepper, and toss to coat evenly. Spread the pieces in a single layer on the pan and roast until golden around the top and sides, about 10 minutes. Flip the pieces over and continue roasting until tender but not too soft, a few minutes longer. Move the pieces in a single layer to a clean sheet pan and leave them to rest and cool.

In a medium sauté pan over medium heat, warm a few tablespoons of oil. Add the apple pieces and sauté, seasoning them with salt and pepper as they cook, until just barely tender. They will overcook and soften quickly, so keep an eye on them. Let them brown a little and then take them off the heat when they are just beginning to turn translucent. The entire cooking time should take only about 3 minutes. Move the pieces in a single layer to a sheet pan and leave them to rest and cool.

Pour oil into a wide medium saucepan until deep enough to submerge the celery root pieces (2 to 3 cups/480 to 720 ml), then heat the oil to 200°F/95°C.

When the oil is ready, add the celery root pieces, immersing them fully. Cook the pieces, rotating and flipping them often, until they are tender when pierced with a fork, 3 to 5 minutes. Using a slotted spoon, move the pieces in a single layer to a sheet pan and sprinkle with salt and some of the lemon juice. Let rest and cool. Reserve the oil.

In a small bowl, drizzle the fennel with a little olive oil and lemon juice, sprinkle with salt and pepper, and toss well. Let rest for about 30 minutes, stirring occasionally.

Before assembling the salad, warm a few tablespoons of the oil from cooking the celery root in a large sauté pan over medium-high heat, add the celery root, squash, and apple pieces, and caramelize for a minute or two to add a little more color. If you like, cut some of or all the wedges into smaller pieces to add some textural variation.

Combine the warm squash, celery root, and apple pieces in a large bowl. Dress with the vinaigrette, adding as much as you like, then add the frisée, spigarello broccoli, fennel, and the figs, if using, and toss everything together, coating all the ingredients evenly with the vinaigrette. Taste and adjust the seasoning as needed, then garnish with the pickled pepper slices.

Traci Des Jardins's Sautéed Artichokes

Knife: chef knife, paring knife

Chef and Owner // Jardinière, The Commissary, Mijita, and Arguello

My first knife, a gift from my aunt, was in a Henckels set that came packed in a briefcase. At seventeen, I walked into my first job wearing a little Laura Ashley dress with my briefcase of knives and everybody laughed their asses off at me, as one of them said, "Whose kid is this, and how did she get in here?"

Artichokes are one of the signature ingredients at Jardinière, and nobody will let me take them off the menu. We use them in everything from bread salad to fish dishes. It feels like I've always prepared them, since that first job in 1983 and throughout my time working in southern France.

It's worth noting that artichokes dull knives quickly and leave a bitter resin on the blade. Always clean off your knives well, *especially* if the blades are carbon steel.

The Cuts

To trim the artichokes, first prepare a small bowl of cold water with a squeeze of lemon juice in it, or have a cut lemon at the ready. Use a paring knife and begin by cutting off the top half—the thicker, leafy part that you would discard if you were eating it whole.

Next, trim away the outer leaves. Hold the artichoke in your secondary hand and, using the springiness of your thumb to push the spine of the knife, turn the artichoke in your hand and move the sharp edge of the blade under the fibrous, outermost layer of leaves.

Next, remove all the green from the stem by continuing to turn the artichoke in your hand. Immediately rub the whole artichoke with the cut lemon or slip it into the lemon water to keep it from oxidizing.

If you are using artichokes that weigh more than ¼ lb/115 g each, quarter each artichoke lengthwise and cut out the choke. For artichokes weighing ¼ lb/115 g or less, halve each artichoke lengthwise and cut out the choke (only if fibrous).

3 Tbsp olive oil

6 medium artichokes, about ¼ lb/115 g each, trimmed (see The Cuts, page 165)

1 fresh rosemary sprig, or 3 or 4 garlic cloves

Salt and freshly cracked black pepper

In a large sauté pan over medium heat, warm the oil. Add the artichokes and the rosemary and sauté the artichokes until they are tender when pierced with a fork, about 4 minutes. Discard the rosemary, season with salt and pepper, and serve.

Jesse Koide's Yakitori

Knife: *honesuki kaku*

Chef // formerly Pink Zebra

My first job was at Ichiban restaurant in San Diego when I was seventeen. I started as a washer, went to prep, then to the back line, followed by the front line, and finally, when I was nineteen, to the sushi counter. My first knives were a 180-mm/7-in *gyuto* and a 23-cm/9-in *yanagi*, both carbon steel.

For yakitori, we separate everything into individual parts and skewer them in any combination. You have to be careful not to shred or otherwise damage the pieces as you cut, and working with the tip of your knife and guiding it precisely by pressing on the face of the blade as you move through helps with that. Using the right knife is important, too. I use a *honesuki kaku*, a Japanese boning knife made especially for poultry, and you should use something similar. When I sharpen it, I go shallow on the *omote* (knuckle) side and a little steeper on the other side. It's a thick knife, but it needs to cut well without letting the sharp edge fold over.

These skewers are delicious grilled plain, but if you like, you can glaze the chicken before or after it is cooked. If I use a glaze, I typically brush only the white meat, which has a little less flavor, but do whatever you like. You can also put the glaze on the table and use it as a dipping sauce.

The Cuts

Begin by separating the chicken's legs from the body. Place the chicken with the legs pointing toward you, breast-side up. Start with the left leg: pull it away from the breast, then score it right down the middle, cutting only the skin with short cuts using the tip of the knife. Using the same technique, continue lifting the thigh away from the chicken's body, and work toward the back of the chicken, scoring the skin until you reach the oyster—it's one of the lumps on either side of the spine. Turn the chicken on its back and let the leg fall on its side so you can work on the pelvis.

To separate the thigh, start at the hip joint. Using the tip of your knife, score along the seam that connects the thigh to the cavity. Once you expose the ball and socket, fold the leg open and crack the ball out of the socket. At the top of the thigh, use the tip of your knife to cut around the oyster, keeping it connected to the back, and cut the thigh away from the body.

Place the chicken on its back and repeat on the right side.

The thighs should be separated from the body. Cut away the belly flaps next. Put the chicken on its back. Poke the tip of your knife in above the ball joint of the hip, and cut along the top edge of the pelvic bone that's pointing towards you and continue to cut the belly flap above it—that hangs down from the breast—away from the bone. To finish cutting away the belly flap, insert the tip of your knife back above the socket joint, but follow the underside of the breast this time to release it.

Now take off the wings. With the chicken on its back, hold the left wing and pull it upward, rolling the chicken onto its side. This will expose where the breast meets the wing. With the wing extended up, use the tip of your knife to separate the wing with short strokes, cutting through the tendons over the armpit first. Cut straight toward the ball joint, trim around the front of the wing, avoiding the breast, and continue around the back side of the wing and around the ball joint. Bend the wing away from the breast, and cut right between the ball and socket of the joint to separate the wing. If it's not too morbid for you, think of it as cutting the arm off by working around the shoulder. You're cutting in a circle around the ball joint, through the tendons that hold the wing close, and once you've snipped those, it'll release easily.

Now remove the breasts. Start with the left side with the breast facing upward and the tail of the chicken pointing toward you. Locate the centerline between the right and left breast, draw an imaginary line where they meet, and score two cuts just on either side of that centerline. Carve along either side of the sternum bone that runs between them toward the tip of the breast. Once the tip of your knife reaches the firm breastplate beneath the breast, you'll want to use the tip of your knife and scoop under the breast and tender to carve it away from the breast plate, gently lifting the breast away as you go. Once you've reached the beginning of the wishbone, where it spreads into a V, use the tip of the knife to cut the breast away from the bone gently, hugging the curve of the wishbone with your knife. Now you'll reach the joint where you pulled off the wing, the only spot where the breast is still connected. Lift the breast away, and cut through that flesh to separate it. Now spin the bird around with the tail facing away from you and repeat on the other side.

The carcass can be used for stocks or broths. The next step is removing the breastplate from the backside of the bird. Find the spot where the back ribs meet the front ribs, and cut right through the cartilage that connects them. Use the tip of your knife to cut through the joints that connect the top and backside of the bird, and pull the top away as you work through. Be firm with your cuts; the bones are tough but cuttable, and there's no reason to be gentle here.

The oyster is located on the back of the chicken above the socket of the hip joint on both sides of the spine, between the hip joint and the back rib cage in a rounded pocket of bone. To take them out, with the skin-side of the chicken facing up, score through the skin and carve a half moon from where the hip joint starts toward the spine back to where the rib cage is. Scoop the tip of your knife under the oyster with small, gentle cuts. Even if you do this sloppily or don't exactly know where the oyster is, you will probably get some of it, and that's fine. Repeat on other side.

After you've removed the oysters, turn the chicken so the tail is facing away from you. Using a similar technique to removing the breasts, create an imaginary line between where you removed the oysters and where the neck would be (or is, depending on where you bought the chicken). Score down that centerline with the tip of your knife. As with the breast, use the tip of the knife to remove the shoulders. Using the tip of the knife, hug the spine with the blade and cut away the shoulder meat from the shoulder blade. Lift the meat away as you trim and slice through to remove it completely. Cut away any excess fat as you go.

You should have two quartered legs, two wings, two oysters, two shoulders, and two breasts.

To cut the wings, find the three distinct pieces: the tip, the midpoint, and the drum. To separate the drum, find the triangle of skin between the drum and the midpoint. Cut along the inside of the drum to where the "elbow" is. Using the heel of the knife and the butt of your hand, make a clean cut through the bones, leaving the joint connected to the drum. Pop through it by pushing down on the back of the blade.

To remove the wingtip from the midwing, locate the joint between them, cut along the inside, and push through it to sever the bone, leaving the joint connected to the tip. Now you have a midwing with no joints on either end. Cut the tip into two pieces, right above where that little thumb piece pokes out.

To butterfly the midwing, locate the two bones running parallel to each other, place the outside of the wing down, and score two lines above the center of each bone. Then use the tip of the knife to trim around the bone, being careful not to cut through the bottom, leaving the bone connected to the meat beneath it. You are exposing the bones but leaving them connected. It helps to hold the knife as a pencil; this gives you more precise lines without using too much force.

To debone the drum, place the skin side of the drum face down. Using the tip of the knife, score a line between the two joints, tracing along the center of the bone. Trim around either side of the bone, carving the meat away from it. You are butterflying the flesh away from the bone as you remove the bone. Once the meat opens away from the bone, lift the bone up and use the tip of your knife with the blade facing outward to gently separate the flesh from the bone, hugging the underside of the bone with the tip of your knife to separate it from the flesh. Then come back inward and cut down right next to where it's connected to the joint to cut the bone away completely.

The tender is naturally separated from the breast; it's the thinner layer laying on top of the breast, connected by some flesh. To separate the breast from the tender, lift up the tender and cut beneath it to sever its connection to the breast. If your tender has a piece of silver skin, peel that off without tearing the flesh. Flip the tender over and locate the thin, white tendon running along the length of it. This is the same tendon you cut through in the armpit. Cut along both sides of it with tip of your knife (a pencil grip is recommended). It disappears at one point into the bottom half of the tender; trace along both sides of it until you can't see it anymore and don't cut through the bottom of the tender. Use the tip of your knife with the blade facing outward to the right and insert it between the tender and the bottom of the tendon. Using a gentle sawing motion, cut it away from the tender. Now come back, cut inward until you reach the point where the tendon disappears, and sever it there.

How you cut the breast for skewers is up to you. You are looking for bite-size pieces that are about ¾ to 1½ in/2 to 4 cm on all sides. To start, cut away the triangle of meat at the narrow end of the breast, about 1½ in/4 cm in length, and cut that lengthwise to make two narrow triangles. Roll them up with the skin on the outside. If the breast feels long, or especially big or thick, cut away the flesh below where the tender was connected and slice it into bite-size pieces. Work through the breast, cutting away strips of similar sizes until you've reached the thickest half, the top half of the breast. If it's greater than 1 in/2.5 cm, cut it into two thinner layers, then slice into bite-size pieces.

To separate the drum from the thigh, turn the leg so the inside is facing up. Hold onto the drum and locate the ball and socket of the knee joint where the drum meets the thigh. Using the whole body of the knife, cut from under the knee (farthest from you) and use a good amount of force to cut through that joint to separate the drum and thigh.

To debone the thigh, trim off any of the excess fat, which you can save for stock (it flares up on the grill). Use the tip of the knife to score along the top center of the bone between the two joints, using your thumb and forefinger to spread the meat away from the bone as you gently carve it away from the bone on either side. Continue until the meat is only connected on the bottom edge of the bone. Now turn it sideways.

Pinch the bone and raise it so the meat hangs down. Tilt the edge of the blade toward the ceiling and draw it along the underside of the bone to cut the meat away from the bone, working away from the kneecap. Repeat until you reach the point where the flesh becomes cartilage and turn the knife down toward the board. Cut straight through to remove the bone and the cartilage from the rest of the leg. As with the breast, if the meat is thick, cut away the smaller muscles. You can find those by pinching and pulling up sections of flesh. Some of them will be distinct muscles that you can cut away and make into their own skewer. If you find tendons, score across them a few times so they don't shrivel up on the grill, or remove them as you did on the tender. Now, just as you did on the breast, cut the meat into consistent, skewerable sizes about 1 or 1½ in/2.5 or 4 cm on all sides.

To debone the drumsticks, hold the drum with the narrower end (the ankle) closest to you and the outside of the drumstick facing down. Try to locate the two muscle groups separated by the bone. Using the bone as a guide and using the tip of the knife, score between those muscle groups. Cut the meat away from the bone, spreading it away from the bone with your thumb and forefinger, until it's only connected along the bottom edge. Turn the leg parallel to you, ankle to the left. Insert the tip of the knife below the bone, tilt it up toward the ceiling, and with a gentle sawing motion, gently separate the flesh from the bone. Work toward the knee, separating the flesh from the bone once you reach the knee. Hold the bone up and come back the other way, lifting the bone up slightly and scraping along the underside with the edge of the knife. Note how all the tendons come together at the ankle; they can be slightly unpleasant to eat, so liberally remove as many of these as you like. Remove any and all cartilage from the remaining flesh (save it for stock). Remove small muscle groups, which you'll recognize as flaps connected to the big, main muscle beneath, by lifting them up and cutting through where they connect to the big muscle. As you did with the other parts, cut into consistent sizes fit for skewering.

To skewer the tender, poke the skewer through one end, then fold the tender back and forth to create an accordion-type skewer. To skewer the breast, fold each piece once, keeping the skin on the outside, and skewer. To skewer the drum, weave the skewer in and out of the flesh to keep the piece butterflied and spread out but do not pierce the skin.

---

1 cup/240 ml sake

2 garlic cloves, smashed

1 thumb-size piece fresh ginger, smashed

1 tsp fennel seeds

1 tsp red pepper flakes

3 green cardamom pods

5 black or white peppercorns

2 dried porcini mushroom slices

2 whole cloves

1 bay leaf

1 star anise pod

1 cup/240 ml mirin

1 cup/240 ml soy sauce

1 small piece kombu (kelp), about 4 in/10 cm square

2 tsp *katsuobushi* (dried bonito flakes)

Honey or sugar for sweetening (optional)

1 whole chicken, butchered for yakitori (see The Cuts, page 168)

To make the glaze, in a small saucepan, combine the sake, garlic, ginger, fennel seeds, pepper flakes, cardamom pods, peppercorns, mushrooms, cloves, bay leaf, and star anise. Measure the depth of the sake in the pan with a skewer and then eyeball the halfway point to that depth. Bring the mixture to a boil over high heat, turn down the heat to a simmer, and cook, skimming any foam that forms on the surface, until the liquid is reduced by half, 15 to 20 minutes. Add the mirin to the pan and again measure the depth of the liquid with the skewer and eyeball the halfway point to that depth. Boil and then simmer the mixture the same way, again cooking until reduced by half. Add the soy sauce, kombu, *katsuobushi*, and a little honey if you want the glaze to have a sweet edge. Again measure the depth of the mixture with the skewer and eyeball the halfway point to that depth. Boil and then simmer the mixture the same way, again cooking until reduced by half. The whole cooking process should take about 45 minutes, and at the end, the mixture should have a good consistency for glazing. Remove the glaze from the heat, strain through a fine-mesh sieve, and let cool.

You can use metal or wooden skewers for the yakitori. If using wooden skewers, immerse them in water to cover for 30 minutes. Prepare a medium-hot fire in a grill. If you wish to mimic a *yaki dai*, remove the grate from the grill, set up two bricks on either side of the firebox on which you can rest the ends of the skewers, and then build the fire in the firebox.

I like to assemble the skewers with the thickest portion of the meat in the middle of the skewer. Always pierce the skin first, then tuck in and go through the flesh. Use only one type of cut on each skewer.

For the wing, which has bones in it, weave the skewer through the skin, meat, and bone, tucking the skin in at the end. Make sure the skin is always drawn over each piece, as it gets much crispier that way. Keep in mind that the breast meat cooks faster than the dark meat, so put a larger piece of light meat next to a smaller piece of dark meat.

As noted in the headnote, the glaze can be used as a glaze or as a dipping sauce. If using as a glaze before cooking, brush the skewers with the glaze, then place them over the fire, either directly on the grill grate or suspended over the fire, in the style of a *yaki dai*. (When dipping, do not use glaze you've used to brush the raw meat.) Grill, turning as needed to color evenly, until the skin is crispy and golden brown and the meat is cooked through. This should take only a few minutes. Serve immediately, with the glaze as a dipping sauce if desired.

NOTE: Before you start, I recommend brining your whole bird overnight. For a 2½-lb/1.2-kg chicken, mix 4 quarts/3.8 L water with 1 cup/280 g salt, ½ cup/100 g of sugar, and if you like, some bay leaves, garlic, and peppercorns. For a bigger chicken, say 3½ lb/1.6 kg, brine it for two nights, and then hang it in your refrigerator for up to seven days. If you're skittish about aging poultry, you can just hang it for a day or two, but the longer you do, the better and more concentrated the flavor will be.

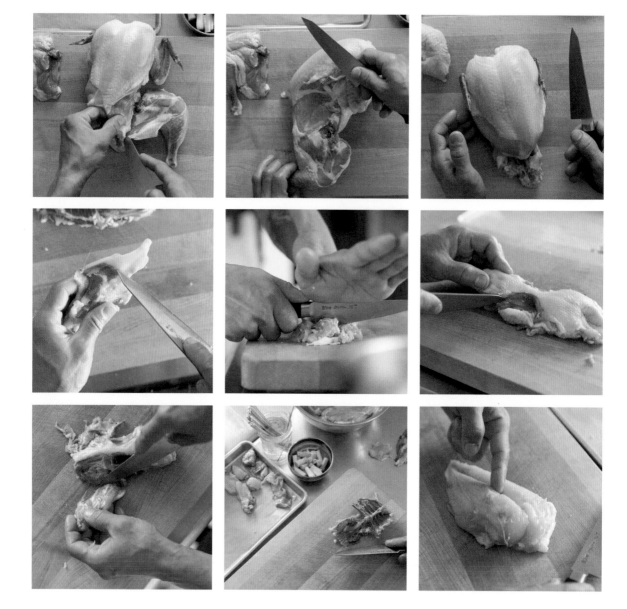

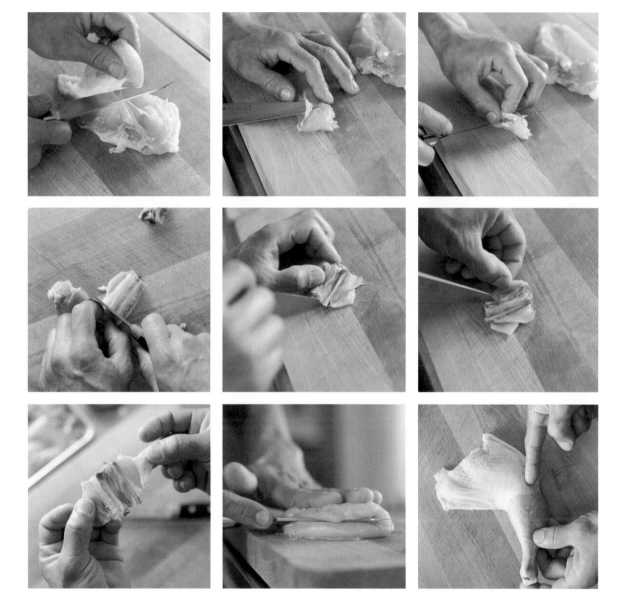

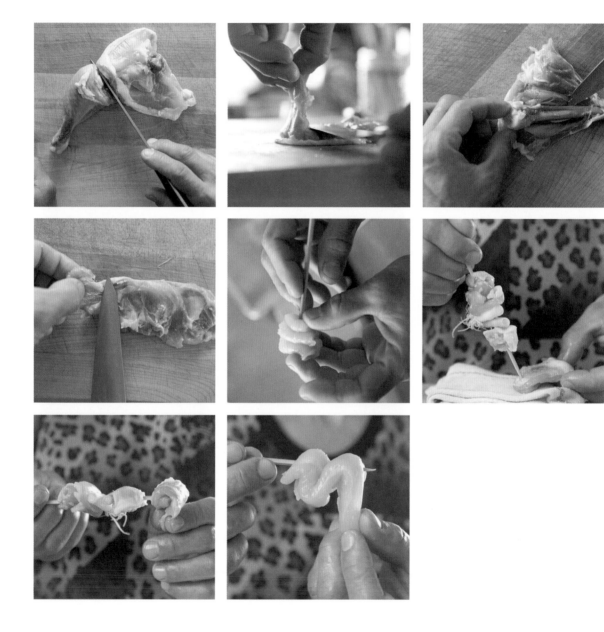

Koichi Ishii's Soba-no-Yamagata with **Vegetables** and **Katsuobushi**

Knife: *soba kiri*, *usuba*, *gyuto*

Soba Chef // Ippuku and Soba Ichi

My first cooking job was at the now long-gone Country Station restaurant, on San Francisco's Mission Street. I used my first knife, a stainless-steel *gyuto*, there.

This style of soba, a vegetable salsa over a nest of noodles, is traditional in Yamagata, my home. The cucumber, *myōga*, and shiso are summer vegetables, so that's when we make it. Although I have given volume measurements for the flour and water for the soba dough, I *highly recommend* weighing both the flour and the water, as it is much easier to be accurate that way.

Here, the noodles are cut from a folded stack of soba dough layers, and a thin, sharp edge is crucial to keeping them intact as you slice. I use a *soba kiri*, designed especially for this task, but you can use any straight, thin, flat-bladed knife with a very sharp, very thin edge (a curved blade will push and distort the noodles). For the vegetables in this recipe, I especially like using a *usuba* for the *negi* (green onion), but for the remaining vegetables and for most vegetable prep in a restaurant, I like the *gyuto* better.

The Cuts

To cut the soba noodles, after your soba dough is folded into a neat stack, grab a straight edge, like a ruler or thin, straight piece of wood, to set up your cuts. Place the guide ⅛ in/4 mm from the edge of the folded dough and place your knife against the guide. To cut, move the knife straight down, pushing through the dough. If you move the knife back and forth in a slicing motion, the noodles will pull apart or press together, and they'll stretch or clump up when you try to cook them.

Move the guide ⅛ in/4 mm in from the new edge and cut again, then continue on in this way across the stack. Ideally, this is done quickly and evenly, but that speed and accuracy takes years of practice. Take your time when you're just beginning, though don't dally too long or the dough will dry out.

Make about twenty-five cuts and place the knife down. Gently pick up the small clump of noodles and shake out the excess flour. Gently bundle the bundles into a small roll and place it in a container with a cover.

To cut the cucumber, cut it in half lengthwise. Scoop out and discard the seeds from each half. Cut each half in half lengthwise. You should now have four long quarters. Holding two quarters together, use a rocking push cut to slice the cucumber crosswise into ½-in/12-mm wedges. Repeat with the remaining two quarters.

To cut the eggplant, trim off the top and blossom end, then cut in half lengthwise. Slice each half lengthwise into batons ¼ in/6 mm wide. Turn the batons parallel to your body, stack them, and slice into ¼-in/6-mm cubes. Soak the eggplant cubes in water to cover for 5 to 10 minutes to remove the *aku*, or "bitterness."

To prepare the green onions, first immerse them in cold water for a minute or so, or rinse them for a few seconds under cold running water. Rinsing mellows out their bitterness, or *aku*, but rinsing for too long will dilute their flavor. Drain, trim off the roots, and, beginning at the bulb end, use a rocking push cut to slice into thin circles.

To cut the ginger, slice off and discard the end, then thinly slice crosswise. Tease the slices apart and then give them a short soak in cold water.

To cut the shiso, stack the leaves with the stems aligned, cut off the stems, and roll up the leaves from stem to tip. Using a rocking push cut, thinly slice crosswise into a fine chiffonade.

Noodles

2¾ cups plus 2 Tbsp/400 grams buckwheat flour (*sobako*)

1 cup plus 1 Tbsp /100 g high-gluten wheat flour (such as bread flour)

Scant 1 cup/220 to 230 g water

Fine buckwheat flour (*uchiko*) for dusting

Although the volume measures appear here, as noted earlier, you will achieve the best results if you weigh both flours and the water. Measure the buckwheat flour and wheat flour, then sift them together into a wide, flat-bottomed bowl and spread the mixture in an even layer in the bottom. Next, weigh 230 g water. Depending on the absorbency of your flour, which varies by season, weight, maker, and type, you might use less. The range is included in the ingredients list to give you a minimum and maximum, but ultimately you'll decide how much to use depending on the texture of the dough as you add it.

Incorporate the water in three or four additions. Slowly add each addition, sprinkling it over the flour and mixing vigorously with your fingertips as you do. You must add the water slowly to make sure both the buckwheat flour and the wheat flour absorb it evenly. Wheat flour tends to absorb the water faster than buckwheat flour does, and if the water is added too quickly, the wheat flour will not share the moisture properly.

As you add the water, the mixture of flour and water will appear dry as it is being mixed, but this is normal. As you work, the dough will start sticking together in small clumps, and you will know that enough water has been added as soon as the dough has picked up the last of the dry flour from the bowl. Keep mixing with your fingertips until the dough is sticky enough to form a ball, then knead the whole mass together vigorously in the bowl, working quickly to keep it from drying out.

To knead the dough, still working in the bowl, press the heels of your hands against the top of the dough, pushing it away from the mass and flattening it smooth. Draw it back in and turn the mass. This will help the gluten develop its strength. The job of the wheat flour in this mix is to provide strength in the noodles, and the gluten develops that strength during this stage of kneading. If the dough is properly hydrated, the surface will look smooth. If the dough is too wet, you'll find the noodles will break up during cooking, and if it is too dry, it will break up when you cut it. You'll know you're done kneading when the dough is soft, smooth, and springy, a little like your earlobe.

Now form the dough into a ball. Fold the dough onto itself once, roll it onto its side, and push down on the side to help force out any air bubbles. This step is important, as air bubbles can cause the noodles to crack.

Dust your work surface with a light sprinkling of fine buckwheat flour and place the dough in the center. Using your palms and working from the center, flatten the dough into a circle by gently pushing outward in every direction. It takes a lot of practice to work the dough into a circle, but starting with it in a ball will help. Remember to work quickly to keep the dough from drying out.

Once the dough is flattened into a wheel, use a *noshibo* or other long, smooth, perfectly straight rod-like rolling pin (no tapering on the ends) to roll out the dough from the center in every direction.

When using the *noshibo* or other rolling pin, your hands must be in constant motion. Start with them together in the center of the roller and push them out toward the edges, stopping the roller just before it meets the edge of the dough. You must not pinch the edges thinner than the rest of the wheel. Using this spreading motion with your hands disperses the flattening over a wide area to ensure both evenness and length. A short, heavy rolling pin would make noodles with irregular thicknesses.

As the circle grows, keep rotating the dough a quarter turn at a time, continuing to flatten it widely and evenly, moving from the center to the edge. Once the wheel is flat and wide—it should be no more than scant $\frac{1}{16}$ in/1.5 mm thick, about the width of a skewer—roll the dough loosely around the *noshibo* or rolling pin and gentle pull the edges toward both ends of the roller. Unroll the dough and roll toward four equidistant corners to shape the wheel into a square.

Sprinkle the square dough sheet with the fine buckwheat flour and fold it in half. Now fold it in half the opposite way, and then fold in half again. You should have a twelve-layered stack. Cut the noodles as directed in The Cuts (page 181), then place in a covered container until ready to cook.

Noodle Broth

Dashi

4 cups/50 g thick-cut dried bonito flakes (*katsuobushi*)

4½ cups/1 L water

Kaeshi

1 cup/240 ml soy sauce

4 tsp mirin

2 Tbsp sugar

Sobatsuyu

3 cups/720 ml dashi

¾ cup/180 ml *kaeshi*

4 cups/960 ml water

To make the dashi, in a medium saucepan, combine the bonito flakes and water and bring to a boil over high heat. Boil the mixture for 10 minutes, skimming off any foam that forms on the surface every few minutes. Then, with the mixture still boiling, use a fine-mesh sieve or other scoop to lift out and discard the bonito flakes. Remove the pan from the heat and measure 3 cups/720 ml to use for the broth. If any remains, reserve it for another use.

To make the *kaeshi*, in a small saucepan over medium heat, combine the soy sauce, mirin, and sugar and warm, stirring often, until the sugar has dissolved. Remove from the heat. You should have a little more than 1 cup/240 ml. Measure ¾ cup/180 ml for the *sobatsuyu*. Reserve the remainder for coating the vegetables.

To make the *sobatsuyu*, stir together the 3 cups/720 ml dashi and ¾ cup/180 ml *kaeshi*.

To finish the noodle broth, stir together the *sobatsuyu* and the water, mixing well. Set aside at room temperature.

Vegetables with Kaeshi

1 piece kombu (kelp), about 5 in/12 cm square, or scant ⅙ oz/5 g natto (minced) kombu

1 cucumber, cut into ½-in/12-mm wedges (see The Cuts, page 181)

1 small Japanese eggplant, cut into ¼-in/6-mm cubes (see The Cuts, page 181)

3 green onions (negi), thinly sliced (see The Cuts, page 181)

4-in/10-cm knob fresh myōga or ginger, thinly sliced (see The Cuts, page 181)

5 fresh shiso leaves, cut into chiffonade (see The Cuts, page 183)

3 Tbsp kaeshi (see recipe page 185)

If using the piece of kombu, place it in a medium bowl with water just to cover. After 15 to 20 minutes, it should be slippery and soft enough to cut. Remove it from the water, reserving the water, and cut the square into strips ⅛ in/3 mm wide. If using natto kombu, submerge it in a little water for a few minutes. The kombu develops a gelatinous texture when soaked that will help the sauce adhere to the vegetables.

In a medium bowl, combine the cucumber, eggplant, green onions, myōga, shiso, and kombu and its soaking water and mix well. Add the kaeshi and stir to coat the vegetables. Let the vegetables stand while you cook the noodles.

To Finish

½ cup/10 g thin-cut dried bonito flakes (katsuobushi)

Noodle Broth (see page 185)

Vegetables with Kaeshi (above)

Bring a large pot filled with water to a boil over high heat. You need a large volume of water—at least 2 to 3 gal/8 to 12 L—to cook the soba properly. If you use too little water, it will cool when you add the noodles, and they won't cook well. Ready a large ice-water bath.

When the water is at a rolling boil, gently spread the noodles in the water and stir with a long chopstick, being careful not to break them. Boil the noodles until they just begin to look clear on the edges and no longer taste of raw flour when one is sampled, 1 to 1½ minutes. Using a spider or other large, long-handled skimmer, lift out the noodles, rinse under cold running water, and then immerse in the ice-water bath and leave until cold. Drain well in a colander or in a bamboo bowl before serving.

To serve, pile a handful of noodles in each bowl and divide the vegetables evenly among the bowls, spooning them on top of the noodles. Add just enough of the room-temperature broth to each bowl to submerge most of the noodles and top each bowl with a big pinch of the bonito flakes.

Chris Kronner's Beef Roulade

Knife: chef knife

Chef and Owner // KronnerBurger

I got my first knife when I was in culinary school: a Wüsthof 10-in chef knife. My first job was at a restaurant called Town Hall. It was as busy as hell, and within three days, I was promoted from intern to pantry cook and found myself prepping everything for the cold station: vegetables, dressings, sauces. At some point, I bought my first Japanese knife, a 240-mm/9½-in Hiromoto carbon-steel chef knife, at Hida Tool in Berkeley. I loved it for a long time, until it was stolen. Then I bought my first Sabatier from Bernal Cutlery around 2008, and I loved how easy it was to sharpen. Even though the edge didn't last as long as the edge on a Japanese knife, it was easy to tune up.

This recipe is a mix of classic roulade recipes from Germany and France, dressed up with my own touches, like bone marrow (I put bone marrow in everything), and in honor of my Sabatier, a few French accents like leeks, herbs, and red wine. Roulade is a great way to end up with a tender, delicious dish from the tough and dry off cuts of the animal.

It's important to have a knife edge with some bite to it for this recipe, as you need to work through rather fibrous sections of meat. The edge can't have too much of a wedge, however. It must be both thin and a bit coarse. That toothy edge helps the knife bite through softer, more yielding cuts during the slicing where a more finely finished edge would simply slide.

The Cuts

To prepare the meat, align the grain parallel to the knife edge. Start your cut on the bottom of the meat, about ½ in/12 mm from the cutting board, and push down with the side of the blade as you smoothly slice back and forth, keeping an even thickness. Press down lightly on the uncut meat with your secondary hand to keep it stable as you go.

Continue slicing back and forth until you get to within ½ in/12 mm of the end, then flip the top portion of the meat over so you can continue to cut along the bottom, proceeding as before. Roll up the thin slice as you go to keep it intact.

Continue this way until the entire piece of meat has been cut into a single long, thin sheet.

To cut the leeks for the roulade, cut off the roots and slice in half lengthwise.

To cut the parsley, gather the sprigs into a small pile and, using the knuckles of your secondary hand to guide the blade slowly over the pile, use a rocking push cut to cut the parsley finely. After the first pass, cut the parsley once again by pinching the end of your blade with your secondary hand and rocking the knife backward and forward over the pile.

To slice the leeks for the braise, cut off the roots, slice in half lengthwise, then use a rocking push cut to slice thinly from the end, using your secondary knuckles to guide the knife.

To cut the celery head, slice off the bottom where the stalks are connected, then place the stalks snugly side by side. Using a rocking push cut, and with your secondary knuckles as a guide, cut off slices ⅛ in/3 mm thick.

To cut the carrot, cut it in half lengthwise. Then, keeping the knife at a slight angle, cut crosswise into half-moons ½ in/12 mm thick.

To slice the garlic, cut off the end, make a few lengthwise slices, and then use a rocking push cut to slice fine pieces crosswise from the end.

To cut the turnips, cut them in half from root to tip, then cut each half in half lengthwise to yield quarters.

To quarter the cabbage, if you're starting with a whole cabbage, stand the cabbage on its root end and cut it in half from top to root, pushing the knife straight down and using the palm of your secondary hand to help push it. Now, with each half facedown, cut each half in half lengthwise. If you are starting with a large cabbage half, place it facedown and cut it lengthwise into quarters.

Roulade

2½ lb/1.2 kg boneless beef with long grain and decent marbling (such as chuck roast, top or bottom round, or beef cheek), in a single piece, cut into a long, thin sheet (see The Cuts, page 189)

2 Tbsp kosher salt

¼ to ⅓ cup/60 to 80 g creamy mustard (such as Dijon or German-style brown mustard)

6 oz/170 g bacon slices

5 small leeks, halved lengthwise (see The Cuts, page 192)

½ cup/20 g fresh flat-leaf parsley, finely chopped (see The Cuts, page 192)

All-purpose flour for dredging

Neutral oil with high smoke point for browning

Unroll the beef horizontally, cut side up, on a work surface. Sprinkle the salt evenly across the surface, then spread with the mustard in an even coat. Lay the bacon slices and leeks along the meat, spacing them somewhat evenly. Sprinkle the parsley evenly over the top.

Cut about ten 12-in/30.5-cm lengths of butcher's twine. Or make the lengths a bit longer if you think your roll will be particularly stout. Now pinch the short left or right side, curl it over, and roll up the meat, working gently to form a tight cylinder. Once you have reached the opposite edge, keeping the seam of the roll on the bottom, loop a length of twine around the roll and secure it on top with an overhand knot. Continue tying lengths of twine around the roll every 1 in/2.5 cm or so, securing them tightly enough to hold the stuffing in place but not so tightly that a bulge develops. If you know how to truss with a single length of twine, you may do that as well.

Dredge the roulade in the flour, shaking off the excess. In a large, heavy, ovenproof saucepan, soup pot, or Dutch oven over medium-high heat, warm a few spoonfuls of oil until hot. To test the temperature of the oil, flick a few drops of water into the pan. If they sputter on contact, the oil is ready. Add the roulade and sear, rotating it as needed to color evenly, until nicely browned on all sides. Carefully transfer the roulade to a plate and set aside. Reserve the pan for preparing the braise.

Braise

5 small leeks, thinly sliced (see The Cuts, page 192)

1 celery head, thinly sliced (see The Cuts, page 192)

1 carrot, sliced into half-moons (see The Cuts, page 192)

4 garlic cloves, thinly sliced (see The Cuts, page 192)

2 large or 3 small turnips, quartered (see The Cuts, page 192)

2 cups/480 ml red wine

1 small or ½ large green cabbage, quartered (see The Cuts, page 192)

3 Tbsp black or yellow mustard seeds

5 lb/2.3 kg beef femur bones, cut into 5-in/12-cm sections

8 cups/2 L beef stock

2 medium celery roots, about 2 lb/900 g each, stems and roots trimmed

About 2 Tbsp unsalted butter

Kosher salt

Preheat the oven to 250°F/120°C.

Put the leeks, celery, carrot, garlic, and turnips in the pan you used to brown the roulade and return to medium heat. Cook the vegetables, stirring occasionally, until lightly caramelized, a few minutes. Transfer the vegetables to a plate. Raise the heat to high, pour in the wine, and deglaze the pan, scraping up any browned bits from the pan bottom. Let the wine simmer for a minute or so.

Return the roulade and caramelized vegetables to the pan and add the cabbage, mustard seeds, and bones. Pour in the stock. Wrap each celery root in aluminum foil with a pat of butter and a sprinkle of salt. Place the pan in the oven and set the celery roots directly on the oven rack next to the pan.

Braise the roulade until it is fork-tender, 4 to 5 hours, depending on the cut you used. The meat is ready when it is springy and tender. It should not be falling apart or at all dry. Remove from the oven and let the roulade rest in the braising liquid for at least 30 minutes. Remove it from the liquid and let it rest again for 5 to 10 minutes before serving. Leave the celery roots wrapped in their foil until serving. While the roulade is resting in the liquid, begin making the roux for the serving sauce (recipe follows). Once the roulade is out of the pan, scoop out and discard the large solids from the braising liquid, strain the liquid through a fine-mesh sieve, and use it to make the serving sauce.

Serving Sauce

½ cup/110 g unsalted butter

½ cup/70 g all-purpose flour

Strained braising liquid from cooking roulade

Kosher salt

1 Tbsp sherry vinegar or red wine vinegar (optional)

To make the roux, in a medium saucepan over medium-high heat, melt the butter. Sift in the flour, a little at a time, whisking continuously until fully incorporated. Continue to whisk as the roux blooms, becoming thinner and smoother. Then as it bubbles, it will swell up. As you continue to whisk, the roux will fall and then bloom once again. The roux is finished after the second bloom, when it is barely browned—it should still be blond—and it smells gently toasted.

Slowly add the liquid to the roux while whisking continuously. Once all the stock is incorporated, continue to whisk until the mixture comes to a simmer and then continue to simmer until the sauce has reduced and thickened enough to coat the back of a spoon. Season the sauce with salt and the vinegar, if you like. Keep warm until serving.

To Finish

Finely chopped fresh flat-leaf parsley (see The Cuts, page 192) for garnish

To serve the roulade, snip and discard the twine and cut crosswise into slices 1 in/2.5 cm thick. Accompany the roulade slices with the warm serving sauce. To serve the celery roots, unwrap them, slice them into wedges, and sprinkle with the parsley.

Taka Tozawa's San Mai Oroshi-Style Filleted Fish

Knife: *deba*

Sharpener // Bernal Cutlery

My first knife was a *gyuto*, which I had when I still lived in Tokyo. I'd gotten a restaurant job when I was fifteen so I could save up to buy my first guitar. Everyone used a *gyuto* in those days because you could do so much with it.

San mai oroshi is a traditional Japanese technique for breaking a fish that emphasizes respect for the animal, and minimal waste. It's important that your *deba* be properly sharpened, as you will be using separate parts of the blade for cutting through fish skin, filleting, and slicing. A rounder edge on the heel of the *deba* toughens it up for breaking through bones, and a flatter edge on the belly is good for slicing. The belly must be sharper and thinner than the heel for fillet work. A finish of between 2,000 and 6,000 grit is good for the *deba* (coarser than for a *yanagi*), so it can bite through the fish skin. This technique works with any decently-sized fish, like a snapper or seabream.

The Cuts

To prepare the fish, begin by scaling if it hasn't been scaled already (many fishmongers will do this for you). Using a fish scaler or the back of your knife—angled at 45 degrees against the direction of the scales—move from tail to head and from spine to belly so you don't miss a spot. Dampening your cutting board with water first will prevent the fish smell from soaking into it.

To break down the fish, first remove the gills. Turn the fish onto its side with the head on your right (if you are right-handed), lift the cheek, and cut the gill membrane by tracing the tip of the knife along the outside of the gill nearest to the body, moving toward the chin. Now turn the fish onto its back so the belly is facing up in the air, and repeat this motion with the opposing gill, moving towards the top of the head. Now follow the chin down to where the collars connect under the mouth, pinch the collars with your secondary hand, and, using a short stabbing motion with the tip of the knife, snap the collar joint under the chin where the collars connect. You don't need a ton of pressure to snap the joint (unlike a bone). Do not twist the knife, as you might chip the blade.

Cut the gills where they connect to the two collars. Open the gill cavity up a little more to expose the spine behind the collars, then sever the gills by cutting them away from the spine, following the contour of the spine in a curving motion. Think of it as shaving the gills off the spine and go lightly.

The belly has two sections: the collar at the fins and the true belly. Keeping the fish on its side and placing the blade at a 45-degree angle to the cutting board, insert the tip of the knife onto the top of the collar inside the cavity and, with short stabbing motions, cut the collars apart, moving down toward the tail. You will be breaking through the collarbones where they meet at the first fins. With a short stabbing cut, and still using the tip of the knife, separate the collars between the fins. Do not cut too deeply into the belly beyond the fins or you will pierce the guts.

Use the belly of the blade instead of the tip. And keep the fish parallel and close to the edge of the board so your knife can get into the correct positions. If the fish is in the center, you won't be able to achieve the necessary angles. Holding the front fin in your secondary hand, slice the belly skin down the center of the fish, cutting down to the anal opening but no further.

Open up the fish and remove the guts by grasping them from the gill area and pulling toward the tail.

Under running cold water, open up the collars to expose the cavity and clean with a bamboo brush, or for smaller fish, with an old toothbrush. You want to remove any remains of the guts and scrub the area around the spine, which has a thin membrane and an air pocket under the membrane. This needs to be scoured and the blood cleaned out with the brush you used for the spine.

To take the head off, place the fish with the head facing left (if you're right-handed) and the belly facing towards you. Slice behind the collar, moving the blade in a straight line at the angle of the collar, stopping short of where the flesh on the other side of the fish begins. Flip the fish over so the spine is closest to you and cut behind the other collar. Line up your knife in the cut you made, and use the heel of the *deba* to cut through the spine, pushing down with your secondary hand if needed to remove the head. Remove the collars from the head by using the heel of the knife to sever the point where they connect with the head. On many fish, the collar (*kama*) is delicious, especially grilled.

To fillet the fish, line up the spine—which runs through the middle of the fish—with the edge of the board closest to you, keeping the tail to the right. Use the belly of the blade to slice along the back (or top, rather) of the fish on the skin only, drawing the blade from tail to head around either side of the dorsal fin. Follow the contour of the fish as you work. When the fish is on its side, the center axis is a little sloped; the head and tail droop down and the spine curves up around the thick middle. It's important to keep the *kiriba* parallel to the spine to cut evenly along the center of the fish, so follow this sloped contour. If the knife angles away, it may cut the fillet below or leave meat on the bones that would otherwise be a part of the fillet.

Return to the cut on the tail side with the tip of the knife and glide the *kiriba* along the bones in the center, slicing toward the head. You should feel or hear little bumps as you ride over the center bones. If you don't hear or feel this, your angle is wrong, and you should try angling the tip a little more steeply toward the body. You do not want to be cutting through the bones here, so if you are about to, lighten up your angle.

The next cut is in the same place but with the tip hitting the spine. Be gentle with the fillet. It's okay to peek in to see your cut, but don't grab the fillet and poke around with your fingers like you would when butchering meat.

Flip the fish around so the belly is facing you and the tail is facing left. Keep the fish on the corner of the board and slice the skin with the belly of the blade, pulling straight toward the tail. Repeat this cut deeper into the fish, and remember to keep the *kiriba* parallel to the board, riding lightly along the bones to remove the flesh, until the tip of the knife touches the spine.

To remove the fillet, flip the fish so the tail is facing to the right, grab the tail (a towel helps here), and with your knife parallel to the board, poke the tip into the fish just above the tail. Push the blade through up to its heel and remove the fillet by gliding the bevel flat across the spine toward the head end. Pull the blade toward you as you cut across the spine to keep the pin bones from ripping through the fillet as they are cut. Cut the attached spot at the tail. If done well, there will be barely any meat left on the bones.

Repeat on the opposite side of the belly. Flip the fish onto the filleted side, with the tail facing right and the belly facing you. Begin to cut at the tail, keeping the kiriba parallel to the board and lifting the end of the belly so you don't mistakenly cut off the head end of the belly.

Flip the fish over, and with the head to the right, begin to cut along the back of the fish going from the head to the tail. Remove this second fillet in the same manner as the first.

To remove the rib bones from the fillet, start with the tail side of the fillet facing you. Use the tip of the knife to cut the pin bones at the beginning of the rib bones. The rib bones extend across the belly on the inside of the fish, and the pin bones extend from the beginning of the rib bones to the sides of the fish in the opposite direction from the ribs. By cutting where the ribs and pin bones are connected before removing the ribs from the inside of the belly fillet, both the ribs and the pin bones are removed more easily.

Use the belly of the blade to cut off the ribs, then use needle-nose tweezers to pull out the pin bones. Remember to pull the pin bones toward the head side of the fillet so you don't rip the meat.

To skin each fillet, set it skin-side down on the cutting board and pinch the skin on the tail side of the fillet with your left hand. With the *ura* side parallel to the cutting board, wiggle the knife while wiggling and pulling the skin. If you do this maneuver well, a little silver layer will remain on the meat. That thin layer often contains flavorful oil, and a skilled sushi chef will leave on as much of it as possible.

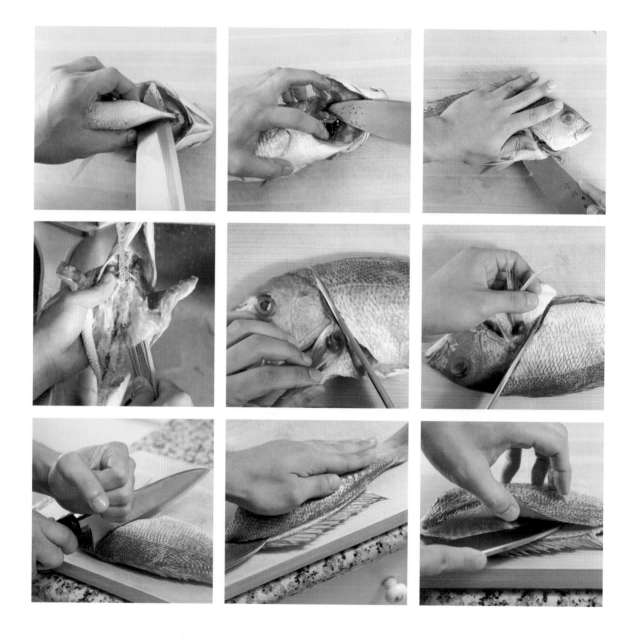

Kelly Kozak's Ramen and Collards

Knife: Chef knife or other wide-bladed knife

Co-Owner // Bernal Cutlery

My first knife was an old Chinese cleaver that I got from a church sidewalk sale in the Tenderloin in 1997. When I'm not helping to run Bernal Cutlery, I work as a food and nutrition educator in San Francisco. I'm often out in the community, offering hungry San Franciscans cooking ideas: what's affordable, nutritious, quick, and can be made in limited kitchen space. My number-one idea usually involves a sharp knife, a cutting board, and inexpensive, nutritious, leafy greens such as collards. Using a simple chiffonade transforms a thick leaf like these into something delicate, making it a perfect partner for noodles (affordable and quick, too). Collards are one of my favorite greens. They're beefy, packed with vitamins, widely available, last a long time in the fridge, and you can trace their origins all the way back to the Fertile Crescent (which just makes for a cool story).

The Cuts

To chiffonade the collards, first de-stem the leaves. Grasp the stem with one hand and, using the other hand, wrap the thumb and forefinger around the base where the leaf begins. Pull along the stem to strip the leaf from the stem. If any leaf remains on the stem, just rip it off. Roll the leaf into a tight cigar, and, using the knuckles of your secondary hand as a guide, slice the roll crosswise into narrow ribbons, about an ⅛ in/4 mm wide.

Ramen and Collards

1 large collard leaf (about 2 ounces /
55 grams)

1 (3 oz/85 g) package Ramen Noodle
Soup

Chiffonade the collards (see The Cuts, page 203).

Cook the noodles following the instructions on the package, using only half
of the flavor packet. Just before serving, stir in the collards so they soften and
wilt. Serve immediately.

Troy Wilcox and Melissa Reitz's Shallots Four Ways

Knife: Petty, paring, or other thin, sharp knife; chef knife

Troy: Cook // formerly Bar Agricole

Melissa: Chef de Cuisine // Locanda

MELISSA: My first job was working at Susanna Foo, a huge Chinese restaurant in Philadelphia where these elderly Chinese men made dumplings all day long, probably for seven cents apiece. Meanwhile, I made fortune cookies and chocolates that were served to diners at the end of the meal. The first knives I had, the ones they give you in culinary school, were crap. But then my chef gave me a MAC vegetable knife when I graduated, and when I took it to the sharpener after a while, he said he'd never seen a good knife so dull. It took me long enough, but after five years, I finally learned how to sharpen it myself.

TROY: My first knife was unremarkable. It was a Wüsthof Trident, and it was in my knife kit in culinary school. Even though it was not a great knife and I found it clunky, it did serve me well. Then I got a Misono, my first Japanese knife, and it was great. It was thin and felt like a cosmic-steel *Star Wars* knife in my hand. Back then, we all went to the hardware stores in Japantown to buy our Japanese knives.

The Cuts

To shuck the oysters, for a good grip and for protection, use a folded towel in your secondary hand to grasp the oyster, with the rounded edge pointing toward the area between your thumb and fingers and the hinged end pointing outward. With the knife in your dominant hand, wedge the tip into the dark spot at the center of the hinge and twist the knife firmly to crack the hinge. Wipe off the knife, then, keeping it angled upward, carefully slide the blade along the inside of the top shell, severing the adductor muscle that holds the shell shut. Neither movement is about strength. Instead, they're about finding the right spot to slip the knife into the shell. Lift off the top shell and discard.

Now carefully slide the knife blade under the oyster meat to detach it from the bottom shell. Leave the oyster resting in the bottom shell.

Shuck all the oysters the same way, changing the towel as needed to keep everything clean and removing any tiny bits of shell and algae from the shucked oysters.

To cut the shallots, cut off the end of the shallot and use a knife and thumb to pull off the outer skin. Trim off the excess root.

Cut the shallot in half lengthwise, leaving the root intact. Place a half cut-side down on a cutting board. Starting at the bottom near the board, cut horizontal slices into the shallot, working from the stem end and stopping just short— about ⅛ in/4 mm—of the root end.

Now use the tip of the knife to make vertical cuts ⅛ in/4 mm thick, working from left to right and stopping just shy of the connected nodule at the bottom of the shallot.

Finally, turn the shallot a quarter turn and, working against the grain, slice vertically ⅛ in/4 mm thick, creating a brunoise cut.

To prepare the anchovies, combine them in a small bowl with water to cover and let soak for 10 minutes. Drain, rinse under cold running water, and carefully fillet them with your fingers. Once filleted, gather them together and use a rocking push cut to chop them coarsely, leaving some irregular pieces.

To finely chop the parsley, first make sure your knife is sharp, as a dull knife will smash and bruise the parsley rather than chop it. Cut off most of the stems, leaving just a small length attached to each sprig. Bunch the parsley into a ball, then, using a rocking push cut, make thin slices, bracing the blade with the knuckles of your secondary hand as you move across the parsley.

To grate the horseradish, peel off the outer skin with a vegetable peeler. Using a Microplane or other fine-rasp grater, finely grate the root. Then, using a chef knife, cut back through the grated root with a rocking push cut. The horseradish must be very fine or it will be tough.

Oysters with Mignonette

This is the classic.

24 oysters in the shell, shucked (see
The Cuts, page 205)

½ cup/120 ml white wine vinegar

½ cup/120 ml dry white wine

¼ cup/35 g shallot brunoise (see
The Cuts, page 208)

1 Tbsp freshly cracked black pepper

Arrange the shucked oysters on a platter and refrigerate until serving.

To make the mignonette, in a small bowl, combine the vinegar, wine, shallot, and pepper and stir well. Let macerate for 15 minutes before serving. You should have about 1¼ cups/300 ml. Invite guests to spoon a little mignonette onto each oyster before eating.

Anchovy Salsa Verde

In this recipe, the vinegar is infused into the shallots. They absorb it, creating a nice little burst of vinegar-shallot flavor in the salsa. This *salsa verde* is excellent with rich, fatty fish, such as salmon or black cod, as well as with beef.

¼ cup/35 g shallot brunoise (see
The Cuts, page 208)

¼ cup/60 ml red wine vinegar

5 high-quality salted anchovies,
soaked in cold water, rinsed, filleted,
and coarsely chopped (see The Cuts,
page 208)

1 bunch fresh flat-leaf parsley, finely
chopped (see The Cuts, page 208)

1 garlic clove, pounded to a paste with a
mortar and pestle

About 1 cup/240 ml flavorful extra-virgin
olive oil

In a small bowl, combine the shallots and vinegar and let macerate for at least 15 minutes or up to a few hours.

In a medium bowl, stir together the anchovies, parsley, garlic, and just enough olive oil to cover everything. Drain the shallots, add to the anchovy mixture, and stir well. Use immediately.

Fresh Horseradish Salsa

When purchasing fresh horseradish root, look for one that is firm and has a smooth surface. Serve this salsa with beef or pork.

¼ cup/35 g shallot brunoise (see The Cuts, page 208)

¼ cup/60 ml white wine vinegar

1 bunch fresh flat-leaf parsley, finely chopped (see The Cuts, page 208)

1 cup/14 g grated fresh horseradish (see The Cuts, page 208)

Pinch of flavorful fine sea salt

Freshly cracked black pepper

About 2 Tbsp extra-virgin olive oil

In a small bowl, combine the shallots and vinegar and let macerate for at least 15 minutes or up to a few hours.

In a medium bowl, stir together the parsley, horseradish, salt, a few grinds of pepper, and just enough oil to cover everything. Drain the shallots, add them to the horseradish mixture, and stir well. Use immediately.

Red Wine Butter

Smear this on a steak and enjoy.

2 cups/340 g shallot brunoise (see The Cuts, page 208)

1 (750-ml) bottle dry red wine

1 lb/455 g flavorful unsalted butter, at room temperature

Big pinch of flavorful fine sea salt

Freshly cracked black pepper

In a medium saucepan over medium heat, combine the shallots and wine and bring to a simmer. Cook until the mixture is almost dry, about 15 minutes.

Drain the shallots, wrap them in cheesecloth, and squeeze well.

Place the butter in a medium bowl. Add the shallots, salt, and several cracks of pepper. Blend with a wooden spoon. Use immediately, or store in an airtight container in the refrigerator for up to 6 months.

Loretta Keller's Calamari Salad with **Marinated Peppers** and **Cucumbers**

Knife: petty, *nakiri*

Chef and Owner // SeaGlass Restaurant at the Exploratorium

My first restaurant job was at a New Orleans fish-fry joint where we didn't even use knives. The only tool I used was a small ice cream scoop to scoop hush puppy batter from a large plastic bucket into a fryer. I left after four weeks.

I've always liked knives—like, really liked them. After my tomboy pocket-knife childhood, I got a small Henckels chef knife for my first serious cooking job. I remember I liked the weight of it but not that it had a plastic handle. Soon after, in the early 1980s, I bought a couple of Sabatier knives in Paris. I still have them.

The thing I like about this dish is the structure of the crunchiness. That's why I marinate the peppers first and add the cucumber at the end. The result is a quartet of bolder flavors, and it's all about distinguishing each one so the diner can experience them individually. The knife work here is especially important to ensure an equal amount of every flavor. In a restaurant, I would use a chef knife with a pushing stroke for most of this knife work, but a petty knife has a lot of benefits for maneuverability, and in reality, it is almost as fast. One great thing about the petty is that it sticks very little when it's used with a pull stroke. I like to use a *nakiri* to cut the onion. When you use it for a pushing stroke, the slices fall effortlessly to the side.

The Cuts

To julienne the basil, sort through the leaves and remove the larger ribs. Stack the leaves with the stems aligned and the largest leaves at the bottom. Roll up the leaves from stem to tip and, using a rocking push cut, slice crosswise into roughly ¼-in/6-mm julienne.

To cut the peppers and chile, use the tip of the sharp, narrow petty blade to remove the seeds and ribs from both peppers and the chile, keeping them intact. Cut the peppers and chile into 3-by-2-in/7.5-by-5-cm strips. Cut the strips into batons ¼ in/6 mm wide. You don't want them to be too thin or they will be limp when cooked.

To slice the onion, lay a whole onion on its side and cut off both ends with the *nakiri*. Using a push stroke, cut in half from end to end, cutting with the grain to keep the sections together. Peel off the skin, cut a half in half again lengthwise, and set aside three-quarters of the onion for another use. Turn the onion quarter on a flat side and, using a push cut, cut into thin, vertical slices. Soak the slices in cold water to cover for about 10 minutes to remove some of the sharp flavor.

To prepare the squid, begin by cleaning them. To clean each squid, separate the head and tentacles from the tube-like body by pulling and wiggling the head, which will come away with the viscera attached. Carefully reach into the body and pull out and discard the long, quill-like cartilage. With your knife, make a cut on the head behind the eyes to remove the tentacles, then discard the head and viscera. Press firmly at the base of the tentacles with your fingers to force out the squid's beak (its small knob of cartilage). When all the squid are cleaned, rinse the tentacles and bodies thoroughly with cold running water. Leave the tentacles whole. Using a pull stroke with the tip of your knife, cut the bodies crosswise into rings ½ in/12 mm wide.

To cut the cucumber, if using a large cucumber, peel it with a vegetable peeler. If using the smaller Persian or other thin-skinned variety, leave the peel on. Cut the cucumber in half lengthwise, place each half round-side down, and scoop out the seeds. Turn the halves flat-side down and, slicing down the width, cut each half into three segments of equal size. Using a pull stroke, cut each segment into strips about the same size as the pepper batons.

2 garlic cloves

Salt and freshly ground black pepper

A handful of fresh basil leaves, cut into julienne (see The Cuts, page 213)

⅓ cup/80 ml red wine vinegar

1 cup/240 ml extra-virgin olive oil

1 red bell pepper, cut into ¼-in/6-mm batons (see The Cuts, page 215)

1 yellow bell pepper, cut into ¼-in/6-mm batons (see The Cuts, page 215)

1 green pasilla chile, cut into ¼-in/6-mm batons (see The Cuts, page 215)

¼ red onion, thinly sliced (see The Cuts, page 215)

2 lb/910 g squid, cleaned and bodies cut into rings ½ in/12 mm wide and tentacles left whole (see The Cuts, page 215)

Neutral oil with a high smoke point for cooking the squid

A splash of red wine or vermouth for deglazing

1 large or 2 small cucumbers (such as Persian or Japanese), cut into ¼-in/6-mm batons (see The Cuts, page 215)

Toasted bread slices or vine-ripened tomato slices for serving

---

To prepare the garlic, put a small amount of salt on a cutting board. (The salt helps prevent the garlic from sliding around and helps to break it down.) Using the side of your knife blade, mash the garlic into a purée, then chop it a bit and smear it with the tip of the knife to create a paste.

To make a vinaigrette, in a medium bowl, whisk together the garlic, half the basil, the vinegar, and the olive oil. Add the bell peppers, chile, and onion and set aside to marinate.

When you are ready to cook the squid, pat the pieces dry and season with salt and pepper. It is best to season the squid right before cooking to prevent the salt from leaching the squid of water. Heat a large sauté pan over high heat until very hot. Add a splash of oil and heat until almost smoking. Add the squid pieces and wait to stir until just after they begin to show some white to pink color, less than 1 minute. Then stir the pieces just until they are white and lightly browned, which will take only 2 to 3 minutes if the squid are not crowded. Immediately transfer the squid to the bowl holding the marinated peppers and onion, suspending the pan over the bowl for about 1 minute to transfer some heat to the marinated vegetables.

Return the pan to the stovetop over medium heat, add the wine, and deglaze the pan, scraping up any browned bits on the pan bottom. Remove from the heat and reserve the wine in the pan.

Let the squid mixture cool to room temperature, add the cucumber, the rest of the basil, and the deglazing wine, and toss gently to mix well. To serve, spoon the squid over the toasted bread.

Sean Thomas's Butter-Poached Salmon with Lentils and Vadouvan Labneh

Knife: utility knife, *gyuto*

Chef de Cuisine // formerly Blue Plate

The ingredients we have in the San Francisco Bay Area often lend themselves to Mediterranean and North African flavors because we share a similar climate with those regions. This dish comes at salmon from a few different culinary angles: the classic French ingredients of fines herbes and butter; a South Asian accent in the *vadouvan*, an Indian-inspired French curry blend; and a distinctive Middle Eastern flavor in the *labneh*, thick, rich strained-yogurt cheese. I know a lot of people who use a *deba* to fillet a salmon, but I'm more comfortable with a *gyuto*. As you work, keep the salmon and your cutting board dry to prevent the fish from slipping, and clean your knife often to make every cut a little easier and more streamlined.

The Cuts

To quarter the shallots, slice them in half from root to tip, then slice each half in half once more, moving with the grain.

To fillet the salmon, position the fish with its belly facing you and the head to the right. Grab the front fin and pull up, then move the knife from heel to tip, slicing diagonally along the gill through the first layer of meat on the right side of the body. It's important to keep your knife at an angle to the line of the spine; otherwise, more force will be needed to move through. Then flip the fish over and make the same cut, but follow through on the spine with the heel of the knife, then come back to that cut line and, using a push cut, press down firmly to take off the head.

To remove the top fillet, lift up the top belly flap and set the knife horizontally, resting it just over the top of the spine. Pressing down with the side of the knife against the spine and keeping the knife parallel to the board, ride over

the spine and cut through all the rib bones where they connect to the spine, moving from head to tail. Position the spine of the knife at a 30-degree angle to the spine of the fish, so as you move it, you are slicing at an angle, not pushing straight through. Move quickly and don't saw back and forth.

To remove the spine, many people flip the fish over at this point, but I leave it on the same side and cut under the spine. Move the knife as you did to remove the top fillet, hugging the bone with the edge of the blade knife and holding the spine with your secondary hand and pulling it to the right as you move. The hardest part is when you get to the back fin. It is difficult to find your way around the fin because the lines are not quite as clear. Don't get too hung up on this step. Instead, just move through. Once you get to the tail, angle the knife down toward the board to finish the cut. A bit of bone is usually left over from the spine and needs to be trimmed off.

To trim the ribs off the belly, angle your knife under the ribs on your right, parallel to the top edge of the fish, and then slice in from the heel to the tip, keeping the blade face at a shallow angle to the board as you move under the ribs. Always cut from head to tail. Hug the ribs with the side of the blade so you don't take off too much belly meat. After the ribs have been removed from the belly, slice off the white flesh of the belly in a clean line.

Take the side of the fish you removed earlier and perform the same motion to cut out the ribs on this side. Again, always move from head to tail.

Now take out the pin bones. Run your fingertips down the length of a fillet, stopping when you feel the tips of the bones. To remove each bone, grasp the tip with needle-nose tweezers and pull it out, sliding it in alignment with its natural orientation. Do not pull it against this path or you will tear the flesh.

If your fish is wide and you have a narrow slicer, use it to slice along the center of the fillet, dividing the fillet into halves along its length. Otherwise, use the *gyuto*. To remove the skin, keeping the knife parallel to the cutting board, slide the knife between the skin and flesh at a 30-degree angle to the long edge, moving smoothly between the skin and the meat. Use your secondary hand to hold the fish gently to keep it from moving during the cut. Next, cut each half of the fillet into three equal portions, for six portions total of ideally around 5 oz/140 g each. Reserve the second fillet of the salmon for another use.

Lentils

2 cups/400 g beluga lentils

8 cups/2 L water

½ carrot, unpeeled

1 celery stalk, halved crosswise

2 shallots, quartered (see The Cuts, page 219)

5 garlic cloves

2 dried bay leaves

½ cup/120 ml extra-virgin olive oil

Kosher salt

Unsalted butter for finishing

In a large saucepan over high heat, combine the lentils, water, carrot, celery, shallots, garlic, and bay leaves and bring to a boil. Lower the heat to a gentle simmer and cook, stirring occasionally, until the lentils are just tender, about 20 minutes. They should be slightly al dente.

Remove from the heat, add the oil, and season with salt. Carefully pour or ladle everything into a separate vessel and let stand for 30 minutes. Remove all the vegetables and the bay leaves and discard. Drain the lentils, reserving some of the liquid for reheating.

Just before serving, in a medium saucepan over medium-low heat, combine the lentils and a few spoonfuls of the cooking liquid and heat, stirring occasionally, until hot. Stir in a little butter to enrich the lentils, then serve.

Vadouvan Labneh

1 cup/240 g *labneh* (strained-yogurt cheese)

¼ cup/25 g *vadouvan* (curry spice mix)

Kosher salt

Whole milk for thinning (optional)

In a small bowl, whisk together the *labneh* and *vadouvan*. Taste and adjust the seasoning with salt if needed. If the *labneh* is too thick for your liking, whisk in milk, 1 tablespoon at a time, to achieve a consistency you like.

Bok Choy

6 baby bok choy
3 qt/2.8 L water

Kosher salt
Extra-virgin olive oil for finishing

Prepare an ice bath. Separate the baby bok choy into individual leaves, removing and discarding any funky-looking parts. In a large saucepan over high heat, bring the water to a boil. Season the water heavily with salt, until it is nearly as salty as seawater.

Drop the bok choy leaves into the boiling water and cook for 20 seconds. Then, using a spider or other long-handled skimmer, lift out the leaves and immediately immerse them in the ice-water bath. When cool, drain well and pat dry.

To serve, preheat the oven to 200°F/95°C. Place the leaves on an ovenproof pan or platter, season with salt, and drizzle with oil. Place in the hot oven to warm for about 1 minute, then serve.

Cranberry Purée

1 cup/140 g dried cranberries
1½ cups/360 ml water

½ cup/120 ml red wine vinegar
½ cup/100 g sugar

In a medium saucepan over medium-high heat, combine the cranberries, water, vinegar, and sugar and bring to a boil, stirring to dissolve the sugar. Lower the heat to a simmer and cook, stirring occasionally, until the cranberries are soft, about 30 minutes.

Remove from the heat, let cool slightly, and then transfer to a blender and purée until smooth. Let cool, cover, and set aside at room temperature until serving.

Salmon

¾ cup/180 ml water

5½ cups/1.2 kg unsalted butter, cold, cut into 1-Tbsp pieces

3 fresh thyme sprigs

3 garlic cloves, smashed with the heel of a knife

1 dried bay leaf

Kosher salt

6 salmon fillet portions, about 5 oz/140 g each (see The Cuts, page 219)

In a wide, heavy, medium saucepan over medium heat, bring the water to a simmer. Add the butter, one piece at a time, whisking continuously until incorporated before adding the next piece. Once you have begun adding the butter, you must whisk the emulsification continuously so it doesn't begin to separate. Also, check the heat of the liquid often by touching it with a fingertip. It should always feel warm to the touch.

As soon as the last bit of butter is fully incorporated, scoop out 1½ cups/360 ml of the emulsified butter for making the *beurre monté* and keep warm. Add the thyme, garlic, bay leaf, and a little salt to the remaining butter, turn down the heat to low, and cook, whisking occasionally, for 10 minutes.

Line a large plate with paper towels and set it near the stove. Working carefully but quickly, add half the salmon pieces to the emulsified butter and poach, still over low heat, just until opaque throughout, about 12 minutes. Using a plastic spatula, gently move each fillet around every couple of minutes to ensure that none of them is sticking to the bottom of the pan.

Using the spatula, carefully remove the salmon pieces from the pan and place them, skinned-side down, on the towel-lined plate to absorb any residual butter. Lightly sprinkle with salt and keep warm until serving. Repeat with the remaining pieces. Strain the butter and reserve for another use.

Fines Herbes Beurre Monté

2 sprigs fresh tarragon

3 sprigs fresh chervil

6 fresh chives

2 sprigs fresh flat-leaf parsley

1½ cups/360 ml warm emulsified butter from poaching the salmon

Kosher salt

Juice of ½ lemon, or about 1 Tbsp white wine vinegar, rice vinegar, or other light-colored vinegar (optional)

Gather the tarragon, chervil, chives, and parsley into a pile and, using a rocking push cut (The Push Cuts, page 152), work from side to side to chop finely.

While the salmon poaches, whisk all the herbs into the warm butter, then taste and adjust the seasoning with salt and lemon juice, if desired. Keep warm until serving.

To Finish

Arrange the salmon pieces skinned-side down and drizzle with the warm *beurre monté*. Accompany with the lentils, *vadouvan labneh*, bok choy, and cranberry purée.

Stuart Brioza's Smoked Duck Breast with Cucumber Salad and Umeboshi-Rosemary Vinaigrette

Knife: *gyuto*

Chef and Co-owner // State Bird Provisions

When I was sixteen, I worked at the Blackhawk Grille in Danville, California. It was 1990, the decade of Wüsthof and Henckels knives, and the first knife I bought myself was an 8-in Henckels chef knife. I still have most of the knives that marked important times in my cooking life. In Chicago, where my wife and business partner, Nicole Krasinski, and I lived almost ten years ago, I bought a set of high-carbon Sabatiers that I still use every day in the restaurant. I even have the same F Dick boning knife that my sous chef at the Blackhawk Grille gave me when I was first learning how to butcher.

The nature of how we cook at State Bird Provisions is to make things a little more natural and a little less square. The oblique cuts in this recipe create an element of imperfection, which is aesthetically nice and provides a combination of soft and crunchy textures in the vegetables. Body positioning is very import-ant when making oblique cuts. It takes a little conditioning to work within a 4-in/10-cm square box, but that's really all the space you need directly in front of your body—there's no overreaching.

The Cuts

To slice the duck, hold your knife at a right angle to the cutting board. Place the tip of your knife on the duck and slice forward through the first half of the breast. Then place your finger over the slice to hold it as you bring the knife back to finish, with the tip resting in its original position.

To cut the cucumber, you want to use a rolling oblique cut. Hold your knife at a right angle to the cutting board, then angle it slightly to one side. Using the first 2 in/5 cm of the blade, cut a chip off the end of the cucumber. Roll the cucumber a quarter turn and make another cut at an oblique angle ½ in/12 mm away. Continue to roll a quarter turn, cut, roll, cut, roll, cut, and so on.

To cut the radish and daikons, roll and cut as you did the cucumber but make the slices slightly thinner, about ¼ in/6 mm at their thickest.

Smoked Duck Breast

2 Tbsp kosher salt

1 Tbsp plus 2 tsp granulated sugar

1 Tbsp freshly ground black pepper

½ tsp pink curing salt #1 (Prague powder #1)

1 boneless duck breast, about 2 lb/910 g, skin and sinew removed

Neutral oil with high smoke point for searing

Place a wire rack on a small sheet pan. In a shallow medium bowl, stir together the kosher salt, sugar, pepper, and curing salt, mixing well. Add the duck breast and turn and toss in the salt mixture until evenly coated. Transfer the breast to the wire rack and discard any mixture remaining in the bowl. Place the sheet pan with the duck in the refrigerator for 2 days.

After 2 days, pull the duck breast out of the refrigerator, wipe off the salt mixture, and pat the breast dry. Let the duck temper at room temperature for 30 minutes.

To cold smoke the duck breast, you'll need to place the breast on a rack above smoldering wood chips. This is easy to do in a backyard grill, if you have one, but any closed space, such as a wooden box or covered fire pit, will work.

In a large stainless-steel or aluminum foil–lined heat-resistant bowl, get about ½ lb/230 g pure hardwood chips (such as alder, oak, or cherry) smoldering. I use a blowtorch, but you can also get them started with a match, or even on your kitchen stove, though you will create a lot of smoke indoors. Once the chips are smoldering, snuff them with some foil and let them smoke. Place the bowl in the bottom of your grill, place the duck breast on the grill rack above the bowl, and then close the grill lid and vents. Leave the duck breast to smoke for about 20 to 30 minutes. You may need to refresh your wood chips two or three times, depending on how well the chips are smoking. To check for doneness, note the smell. Once the duck smells somewhat strongly of smoke, you're done.

When the duck breast is ready, carry it into the kitchen. About 20 minutes before you are ready to serve, place a sauté pan over medium-high heat until hot. Add just enough oil to film the pan bottom and heat until the oil is hot. Add the duck breast and sear, turning once, until the exterior is a golden bronze, or the interior reaches about 135°F, about 2 minutes on each side. Let the duck breast rest at room temperature for 10 minutes before slicing.

Umeboshi-Rosemary Vinaigrette

½ cup/120 ml rosemary oil

½ cup/120 ml dashi

¼ cup/45 g *umeboshi* (salted plums)

½-in/12-mm square fresh ginger, peeled and grated on a Microplane or other fine-rasp grater

1 garlic clove, grated on a Microplane or other fine-rasp grater

In a small bowl, whisk together the oil, dashi, *umeboshi*, ginger, and garlic, mixing well. Set aside at room temperature until needed. You will likely have some left over; store in an airtight container in the refrigerator for up to a few days.

To Finish

1 Smoked Duck Breast (recipe page 228)

2 Tbsp grapeseed oil

1 head maitake mushroom, about 5 oz/
140 g, tough stems removed, torn into
2-in/5-cm pieces

Kosher salt

1 Japanese cucumber, oblique cut
(see The Cuts, page 226)

1 watermelon radish, oblique cut
(see The Cuts, page 228)

1 green daikon, oblique cut
(see The Cuts, page 228)

1 white daikon, oblique cut
(see The Cuts, page 228)

1 cup/240 ml Umeboshi-Rosemary
Vinaigrette (see recipe page 229)

2 Tbsp shiso microgreens

2 Tbsp black sesame salt

Slice the smoked duck breast (see The Cuts, page 226). You should have about 20 slices. Set aside.

In a large sauté pan over medium heat, warm the oil until small wisps of smoke begin to rise from the pan. Add the mushroom pieces, season with salt, and cook, stirring occasionally, until golden brown on all sides, about 5 minutes. Remove from the heat.

In a large bowl, combine the cucumber, radish, daikons, and mushroom. Drizzle with the vinaigrette and toss until evenly coated.

To assemble, spoon the cucumber salad onto a large platter or divide evenly among five individual plates. Arrange the duck slices around the salad, then garnish with the shiso and the black sesame salt. Serve right away.

Nick Balla's Knife-Cut Cold Ginger Noodles with Sunflower Seven Spice

Knife: chef knife, specialized noodle-cutting knife (optional)

Executive Chef // Duna

My first job was as a dishwasher at a wild game–themed restaurant in Michigan, where I worked up to chef after a few years. My first knife was a 9-in Wüsthof chef knife, and I still love it. It's given me a lot of amazing cuts over the years.

This was a favorite dish of the kitchen crew at Motze, our project before Duna. Harry Kongvongsay, my sous chef, is cutting the noodles here. We cook what we want to eat, and this recipe, which draws on ingredients typically in our larder, is our interpretation of the classic cold sesame or peanut noodles found on Chinese restaurant menus. It's a relatively simple dish, and the cuts are designed to work in tandem with the shape and texture of the noodles.

This recipe calls for ginger starch, which is a by-product of juicing fresh ginger. At the restaurant, we juice the ginger and transfer it to a container, where the starch gradually settles to the bottom. We pour off the juice and then we dry the starch in a dehydrator. We have found the starch gives the noodles an interesting texture when they are cooked. If you don't have ginger starch, you can substitute an equal amount of potato starch and add 1 teaspoon ground ginger to it. The braised kombu in this dish takes more than an hour, most of which is idle time, so I suggest you start simmering it before you begin the noodles, and keep an eye on it as you work.

At Motze, we made most of our ingredients in-house, including the seven-spice mix that tops the noodles just before serving. But if you are pressed for time, you can swap in store-bought *shichimi togarashi* (seven-flavor chile pepper) for the spice mix. Look for it at specialty Japanese markets and well-stocked supermarkets. You can hunt down the orange peel powder and green garlic powder at a specialty retailer online, like La Boîte or Le Sanctuaire, or use the recommended substitutions.

The Cuts

To cut the noodles, using a heavy, sharp, flat-bladed knife like a chef knife or a specialized noodle-cutting knife, use a down push cut, using your secondary hand to guide the knife ¼ in/6 mm at a time. Do not use a slicing motion or move the knife back and forth. Tease the noodles apart after cutting.

To julienne the kombu after braising, fold it in half lengthwise and, using a rocking push cut, cut into thin julienne strips from the end.

To cut the green onions, first slice off the root end. Slice a palm-size length from the white portion of the stalk, then drag the tip of your knife down its length, cutting only to the core, not all the way through. The core may be a concentric circle or twinned cores, and if the onion is beginning to flower, it will be less uniform. If the core seems tender, cut it into julienne: flatten the outer layer of the green onion and, using a rocking push cut, cut lengthwise with the grain into julienne strips. If the core seems a bit tough, cut it into rings against the grain. A sharp knife is especially important when cutting green onions.

Braised Kombu

1 cup sake	1 cup honey
1 cup soy sauce	2 pieces kombu (kelp), each about 4 by 6 in/10 by 15 cm

In a small or medium saucepan over medium heat, combine the sake, soy sauce, and honey and bring to a simmer. Add the kombu, reduce the heat to very low, and simmer for 45 minutes.

Remove the kombu from the pan and julienne it (see The Cuts, above). Return it to the pan and continue to simmer over very low heat until al dente, about 30 minutes or so.

Remove from the heat and pour through a fine-mesh sieve placed over a heatproof bowl. Set the kombu strips aside for serving. Let the liquid cool to room temperature, then store in an airtight container in the refrigerator for up to 1 month. Use the liquid to glaze anything from pork and fish to chicken and vegetables.

Ginger Noodles

2 cups/280 g all-purpose flour

¼ cup/55 g ginger starch, or ¼ cup/55 g potato starch plus 1 tsp ground ginger

2 large eggs, lightly beaten

2½ Tbsp water

1 Tbsp salt

In a medium bowl, combine the flour, ginger starch, eggs, water, and salt and stir with a wooden spoon until the dough comes together. Cover the bowl with a plate to keep the dough from drying out and let rest on a countertop for 15 minutes.

Uncover the bowl and knead the dough in the bowl for 5 minutes. Re-cover the bowl and let the dough rest again for 15 minutes. Repeat this process three more times to develop the gluten in the dough.

During the final resting period before the final round of kneading, put a large pot filled with water on the stove over high heat. Then liberally flour a work surface.

Once you've finished kneading the dough, cut the dough in half and shape each half into a thick rectangle. Place a rectangle on the floured surface, flatten it with the heels of your hands, and then roll it out into an 18-by-12-in/45-by-30.5-cm rectangle about ¼ in/6 mm thick. Lightly flour the top of the dough sheet, then fold it into thirds as if folding a business letter. Flip the dough over so the seam is on the bottom and a long side is facing you. Cut the dough crosswise into noodles ¼ in/6 mm wide (see The Cuts, page 235). Repeat with the second rectangle of dough.

At this point, the pot of water should be boiling. Immediately transfer half the noodles to the boiling water and boil for about 3 minutes. The noodles should be chewy but cooked through. Using a spider or other large, long-handled skimmer, lift out the noodles and rinse under cold running water until cool. Drain the noodles well, then wrap them in a towel so they do not absorb extra water and refrigerate until ready to serve. Repeat with the remaining noodles.

To serve, transfer noodles into a large serving bowl. Toss with the sunflower tahini sauce to coat. Garnish with the braised kombu and seven-spice mixture.

Sunflower Tahini Sauce

1 cup/140 g toasted sunflower seeds

2 Tbsp honey

1 Tbsp fresh lime juice

1 Tbsp cider vinegar

1 cup chopped fresh cilantro

3 garlic cloves

1 serrano chile, stemmed

1 Tbsp fine sea salt

Apple juice or dashi for thinning (optional)

In a high-speed blender, purée the sunflower seeds until completely smooth. Add the honey, lime juice, vinegar, cilantro, garlic, chile, and salt and again purée until completely smooth. You should have about 1 cup/240 ml. Use immediately, or transfer to an airtight container and refrigerate for up to 2 weeks. If your sauce is too thick—the consistency of nut butter or thicker—add up to ½ cup/120 ml apple juice or dashi to thin. The consistency should be thick enough to coat the back of a spoon and drip off, not stream off.

Sunflower Seven Spice

¼ cup/40 g crushed toasted sunflower seeds

2 Tbsp toasted whole sunflower seeds

1 Tbsp orange peel powder or grated fresh orange zest

4 dried árbol chiles, toasted in a dry pan until fragrant and then crushed

1 Tbsp green garlic powder, or 1 tsp conventional garlic powder

2 Tbsp crushed toasted wild nori

2 Tbsp coarse tomato powder

1 Tbsp toasted black sesame seeds

1 Tbsp flake sea salt

In a small bowl, stir together the crushed and whole sunflower seeds, orange peel powder, chiles, garlic powder, nori, tomato powder, black sesame seeds, and salt, mixing well. You should have about 1¼ cups/190 g. Use as needed for the noodles, then transfer the remainder to an airtight container and store in a cool, dry pantry for up to 1 month.

Tim Ferron's Beef Jerky

Knife: *sujihiki*, Western scimitar, or other thin slicer

Sharpener // Bernal Cutlery

My first job was as a dishwasher at Disneyland. Next, I was a fry cook in an airport hangar–themed restaurant, prepping the burger station and working eight fryers in a row. Then I worked at a fine-dining restaurant prepping fifty tenderloins at a time in a giant tub filled with gallons of this and that. At some point, I realized that cooking wasn't intimate anymore, so I complained to my neighbor, and she hooked me up with a job at Avedano's Meats, in San Francisco's Bernal Heights.

You can use any lean, large cut of meat for jerky. I like eye of round or top round. I also like to wrap it in cellophane and pop it in the freezer for about 1½ hours, so it will keep its shape a little better when I cut it. Because the beef toughens a little as it dries in the oven, the tenderness of the cut you choose for making jerky is important. Regardless of the cut, however, remember always to slice against the grain. Meat is similar in structure to wood (and really, to lots of organic materials). Cutting against the grain severs the muscle tissue and gives you shorter fibers that are easier to chew. If you cut with the grain, you leave whole muscle tissues intact, resulting in a very tough jerky that is hard to rip apart. And there is nothing you do with jerky but chew.

The meat needs to marinate for 8 to 12 hours, so plan your day accordingly.

The Cuts

To cut the ginger, peel the skin with a spoon. Using a pull stroke, cut broad, thin slices from the long side of the root. Stack the slices and, using a pull stroke again, cut the pieces into long, narrow matchsticks.

To zest the lemon, rub the skin against a Microplane or other fine-rasp grater, being careful not to remove the bitter white pith with the colored zest.

To cut the beef, using a pull stroke and moving against the grain, slice the beef into thin sheets about ⅛ in/4 mm thick.

---

1 tea bag strong black tea

4 cups/960 ml boiling water

¾ cup/180 ml whiskey

2 Tbsp honey

2 Tbsp matchstick-cut fresh ginger (see The Cuts, above)

Grated zest of ½ lemon (see The Cuts, above)

1 (3-in/7.5-cm) cinnamon stick, broken into a few pieces

7 or 8 whole cloves

1½ Tbsp kosher salt

4 lb/1.8 kg boneless lean beef, in a single piece, thinly sliced (see The Cuts, above)

To make the marinade, place the tea bag in the boiling water and let steep until the tea becomes a little bitter, 5 to 10 minutes. Meanwhile, in a large bowl, combine the whiskey, honey, ginger, lemon zest, cinnamon, cloves, and salt and mix well.

When the tea is ready, discard the tea bag and add the tea to the other ingredients, stirring until the honey and salt are dissolved and all the ingredients are well mixed.

Add the beef slices to the marinade, submerging them fully. Cover and refrigerate for 8 to 12 hours.

Preheat the oven to 150°F/65°C or to its lowest setting. (If the lowest setting is higher than 150°F/65°C, prop the oven door open with a wine cork.)

Drain the beef and pick off any stray pieces of ginger or other seasonings. Pierce the end of each slice of beef once with a toothpick, then hang the slices on the bars of the oven rack, using the toothpicks as supports and turning the slices perpendicular to the bars. Make sure the slices are not touching one another, and put a sheet pan in the bottom of the oven to catch the drips. Close the oven door and set the timer for 3 hours.

When the timer goes off, check the jerky. If the meat is leathery and firm but malleable, it is ready. If it is still soft and wet, leave it in the oven longer and set the timer for another 30 minutes. If it is crumbling or cracking, you've left it in the oven too long.

Let the jerky cool completely before eating. It will keep in an airtight container at room temperature for up to a few weeks or vacuum sealed for a few months.

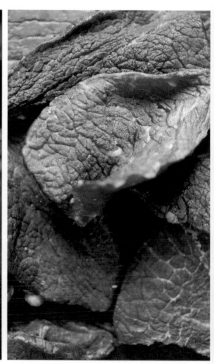

Lisa Weiss's Hawaiian Chili and Liliko'i Bang-Bang

Knife: *nakiri*

Sharpener // Bernal Cutlery

My first industry job was working at a kava-centric health food café in Diamond Head on Oahu. It was the start of the açaí bowl craze, and a lot of tourists were coming in on Segway tours for "healthy" smoothie-in-a-bowl stuff. I dimly recall using house Dexter serrated knives for everything, from cutting up fruit for smoothies to making hummus wraps. Absolutely no meat sauce here, kids.

I love salty meats, tropical juices, and buttery coconut flavors. These are my happy flavors, and you can find them all in a cup from Zippy's, Hawaii's home-grown restaurant chain. I grew up seeing Zippy's Hawaiian chili at school fundraisers, birthdays, and funerals and probably ate it at midnight on McCully Street after a few bong hits (but who can remember?). Despite the general beigeness and downright pedestrian concept of chili, Zippy's is actually quite amazing. And then there was Zippy's "Orange Bang"— a somewhat baffling accompaniment to the chili. It is the heavy metal of meal pairings. I use *liliko'i* in my version. If navel orange is a Duck Tour ride, then *liliko'i* is a car chase. And why do I call it a bang-bang? I'll leave it to *Louie*, season four, episode three to explain that.

When you pick a pineapple for this recipe, don't pick out one that's too green or too yellow. If you hold it up and sniff it, it should smell like a pineapple. Look out for mold on the cut-off stem, too. A lot of recipes for Hawaiian call for canned pineapple, but the fresh stuff has better flavor; more tang, and the flavor sweetens as it cooks. I like using a *nakiri* for bigger fruits and vegetables, and push cuts work better than pull cuts with the big stuff. For smaller work, the pull cut is great for extra control.

The Cuts

To cut the onion, use a push cut to slice off the stem, then turn the onion onto the flat end. Cut the onion in half from root to stem end. Using a push cut here offers more force and control than a pull cut. Peel each onion half. Lay an onion half on its flat side, cut it horizontally into three or four even layers and then cut it vertically into three or four even layers. Use a rocking push cut to slice it vertically into ¼-in/6-mm dice.

To grate the ginger, I like to use an *oroshigane* (Japanese grater) because it breaks up the ginger into a paste and you're left without the fibers. Rub the ginger in a small circle on the *oroshigane*, then use a bamboo brush to remove the ginger from the grater. If you don't have the customary bamboo brush, rub the grated ginger off with the nub of ginger. Don't try to grind the ginger nub to its very end, as the *oroshigane* is sharp and you will destroy your fingers. Alternatively, use a fine-rasp cheese grater or a Microplane grater to grate the ginger.

To dice the carrot, first peel it, then cut off the ends. Segment it widthwise into three parts, then cut a thin plank off one side so you have a flat face you can rest on the board. Use a rocking push cut to slice ¼-in/6-mm or thinner planks, lengthwise.

Stack the planks again so that there's a flat face resting on your board. Then use a pull cut to slice match sticks lengthwise. Gather a manageable bundle of matchsticks and use a push cut to slice off nice, square dice (or brunoise).

To cut the pineapple, lay it on its side and cut off the top (which you can plant and grow!) and then cut off the bottom. Next, stand the pineapple on one of its flat ends and, following the contour of the fruit, cut downward to remove the skin, cutting off as little of the fruit as possible. Even though it seems wasteful, you must cut deeply enough to remove the spiny eyes. Now cut straight down the middle of the pineapple along its center axis (a push cut is definitely best here). Rotate the pineapple a quarter turn and cut straight down again, to create quarters. Finally, cut downward along each quarter to remove the core portion. Although the core can be eaten, it is quite fibrous and will turn into a chewy wad in the chili.

Turn each quarter onto its flat, cored side and, using a pull stroke, cut the pineapple crosswise into planks ¼ inch/6 mm thick. Cut the planks into ¼-in/6-mm cubes.

To dice the bell pepper, turn it on its side and cut off the stem. Cut the pepper in half lengthwise, pull out the seeds, and trim away the pith. Cut the halves lengthwise into slices ¼ in/6 mm wide. Arrange the slices together and, using a push cut, slice into ¼-in/6-mm dice.

To mince the chives, line them up lengthwise and use a rocking push cut to slice into very small pieces.

Chili

1 small pineapple, peeled, cored, and half diced (see The Cuts, above), or 1 (12-oz/340 g) can crushed pineapple

1 lb/455 g ground beef

1 small yellow onion, diced (see The Cuts, page 247)

1 Tbsp grated fresh ginger (see The Cuts, page 247)

1 large carrot, diced (see The Cuts, page 247)

2 tablespoons soy sauce or tamari

¼ cup/30 g chili powder

2 Tbsp salt

1 tsp ground cumin

½ tsp freshly ground black pepper

½ tsp cayenne pepper

1 Tbsp Gochujang (Korean red pepper paste)

1 bell pepper, diced (see The Cuts, above)

1½ cups/360 g canned crushed tomatoes

½ cup/80 g dried kidney beans, soaked in water to cover overnight and drained, or 1 (12-oz/340-g) can kidney beans, drained and rinsed

½ cup/90 g dried pinto beans, soaked in water to cover overnight and drained, or 1 (12-oz/340-g) can pinto beans, drained and rinsed

3 cups/720 ml beef stock

1 bay leaf, preferably fresh

Cooked sushi rice or brown rice for serving

Grated sharp Cheddar or Monterey Jack cheese for garnish

1 bunch chives, minced, for garnish

Add half the pineapple cubes to a food processor or high-speed blender and pulverize until somewhat smooth. Measure 1 cup/240 ml of the smooth pineapple and pour it into a medium bowl. Add the remaining cubed pineapple to the bowl and stir to mix well. You will probably have some smooth pineapple left over, which you may want to add to the final chili if you feel it needs more pineapple.

Place a large pot over high heat, add the beef, onion, ginger, carrot, and soy sauce, and stir for a couple of minutes to mix well and to break up the meat. Add the chili powder, salt, cumin, black pepper, cayenne pepper, and red pepper flakes and mix well. Then add the pineapple and the bell pepper, stirring frequently until the beef is browned. Add the tomatoes, kidney and pinto beans, stock, and bay leaf, bring to a boil, and then lower the heat to a simmer. Cook, uncovered, stirring occasionally and skimming the fat from the surface every so often as needed, until the liquid reduces and the mixture thickens, about 2 hours.

Taste and adjust the seasoning with salt and black pepper and with any leftover pineapple if needed. Serve with a side of rice and throw some grated cheese and chives on top of dat faka!

Liliko'i Bang-Bang

16 satsuma mandarins, peeled and sectioned

4 *liliko'i* (passion fruits), halved crosswise

3½ cups/840 ml light coconut milk

2 tsp vanilla extract

Ice for serving

Place a food mill over a large bowl. Toss in the satsuma sections and crank away to separate the juice from the fibers and pulp. Next, scoop the seeds and gooey pulp from the *liliko'i* halves into the mill and crank away to separate the juice from the seeds. (If you put the *liliko'i* in before the satsumas, you may end up with seed shavings in your juice and it will taste a little bitter.)

If you don't have a food mill, you can still make the bang-bang. Cut each satsuma in half crosswise and use a hand juicer to extract the juice. Heap the seeds and pulp from the *liliko'i* halves into a fine-mesh sieve placed over a bowl and spread them around with a whisk until the juice passes into the bowl.

Add the coconut milk and vanilla to the fruit juices and whisk together until frothy. Pour over ice in tall glasses. Enjoy with chili and with any heavy meal riff for maximum bang-bang satisfaction.

Index